IMAGES
of America

EDISON FIREFIGHTING

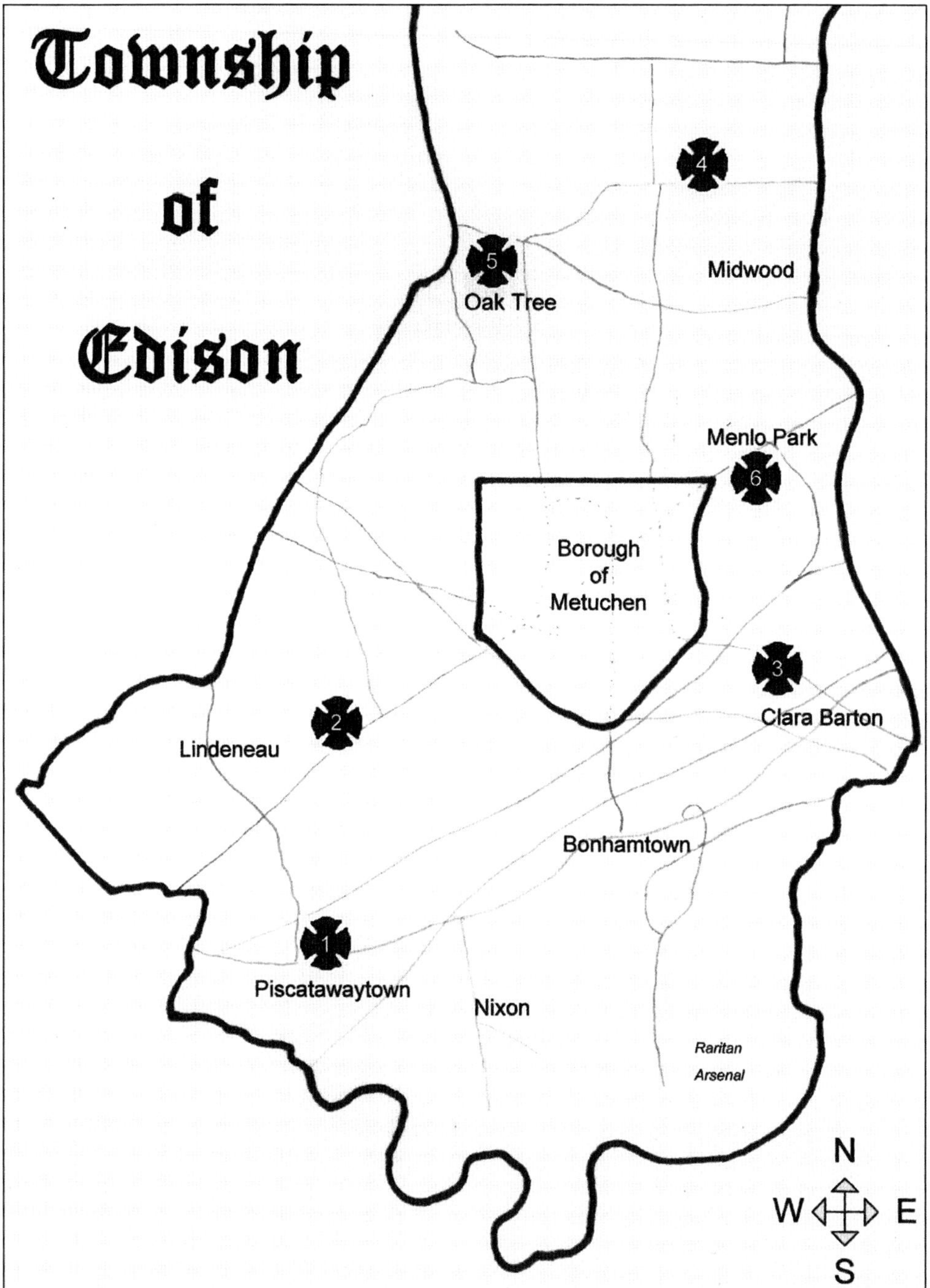

Shown here is a map of Edison Township with its subsections. The Edison Division of Fire has six firehouses with each location indicated by the station number within a Maltese cross.

On the cover: Firefighters are combating a fire that ripped through Edison's landmark catering hall, the Pines Manor, on February 2, 1972. (Courtesy of Edison Division of Fire, photograph by George Smith.)

IMAGES
of America

EDISON FIREFIGHTING

Eugene A. Enfield Jr.

ARCADIA
PUBLISHING

ISBN 978-0-7385-5491-4

Published by Arcadia Publishing
Charleston SC, Chicago IL, Portsmouth NH, San Francisco CA

Printed in the United States of America

Library of Congress Catalog Card Number: 2007927176

For all general information contact Arcadia Publishing at:
Telephone 843-853-2070
Fax 843-853-0044
E-mail sales@arcadiapublishing.com
For customer service and orders:
Toll-Free 1-888-313-2665

Visit us on the Internet at www.arcadiapublishing.com

This book is dedicated to all Edison firefighters, past and present, who demonstrate their selfless dedication to protecting the community, and for their loved ones who pray for their safe return.

Contents

ACKNOWLEDGMENTS

For the past six months, I have scanned over 400 photographs and interviewed many people throughout Edison Township. Needless to say, I could not have completed this project without their assistance. I would like to express my deepest appreciation and acknowledge the following people for their help: all of the members of the Edison Division of Fire, especially acting fire chief Norman Jensen; Deputy Chief Ralph Ambrosio Jr.; Captains Brian Latham, Eugene Whitecavage, James Hook, David Davis, Robert Sofield, and Robert LaCour; Lieutenants Thomas Walsh, Anthony Landi III, Joseph Szebenyi III, and Joseph Horvath; and firefighters Brian Daugherty, John Lindquist Jr., Richard Aszman, Scott McCloughan, Duane Clause, Douglas Kosup, Patrick J. Novia, Robert McDonnell, Edward Hummel, Joseph Anselmo, Daniel DiNuzzo, Andrew Toth, Frank Eosso, Wayne Enoch, and Joseph Toth for providing me with photographs, information, and assistance.

I had the pleasure of meeting several retired firefighters who provided me with information from the "good old days." I extend my deepest gratitude to the following retirees: Chiefs H. Ray Vliet, Richard Latham, Albert Lamkie, and G. Robert Campbell; Deputy Chief Donald Dudics; Captains Robert Grand-Jean, William Schneider, Walter Ulrich, and Robert Coleman; Lieutenants Patrick Novia, Donald Freeman, and Richard Campbell; Inspector Joseph Szebenyi Jr.; and firefighters Leon Mazur, Emery Tibok, and Robert Meany.

I would like to thank the following volunteer fire chiefs for sharing their company's photographs and information: Raritan Engine Company No. 1 acting chief Henry Maurath, Edison Volunteer Fire Company No. 1 chief Walter Boychick, Raritan Engine Company No. 2 chief Robert Barrett, H. K. Volunteer Fire Company chief John Pencak, and Oak Tree volunteer chief Kenneth DiFrenza.

Former volunteer chiefs Morris Kaplan, Lee Sofield, Mitchell Erceg, Ludwig Roell, Louis Zavis, William Prairie, Theodore Wolan, and John Peters provided photographs and helped "fill in the gaps"—thank you for all your help.

In addition, I would like to thank my mother and father for all of their support; the Metuchen-Edison Historical Society and its president, Walter Stochel Jr.; Kris Aszman from Kris Krupp Photography; Patricia and Mitch Ruben from P.M. Studios; Sentinel's photo editor Thomas Caiazza; and Karen Daugherty, Jim D'Amico, Robert Kearstan, Salvator Quagliariello, Elaine Kilijanski, Norman Shamy, Marion Cherick, Teresa Asprocolas, Joan Coleman, and Helen Kosup for their assistance.

Finally, and most importantly, I extend my deepest gratitude to my wife, Grace, and my two boys, Alex and Andrew, for all their support and encouragement. I am extremely grateful for all your help!

Introduction

The purpose of this book is to showcase almost 100 years of historical events and information about Edison's fire service from the early days of bucket brigades and horse-drawn wagons to today's dynamic and complex world of firefighting. I hope that you will be able to appreciate how this fire department has grown from a few inspired volunteers setting out to protect their community into a strong unit of passionate firefighters valiantly serving the fifth-largest municipality in New Jersey. The presentation of familiar events, places, names, and faces in this book exhibits how the tight brotherhood of firefighting exists not only because of its deep and colorful history, but because of its intricate connection with the community that it serves and because of the support of family and friends.

In the early 20th century, Raritan Township (known today as Edison) was a rural town with a small population of approximately 2,000 people. Areas known as Piscatawaytown, Nixon, Bonhamtown, Clara Barton, Menlo Park, and Oak Tree divided Raritan into separate sections. Today over 100,000 residents live within the 32-square-mile township of Edison, which was ranked 28th in the nation in 2006 as one of *Money* magazine's best places to live.

Edison's fire service originated as five separate volunteer fire companies: Raritan Engine Company No. 1, Edison Volunteer Fire Company No. 1 (informally known as Menlo Park Fire Company), Raritan Engine Company No. 2, H. K. Volunteer Fire Company, and Oak Tree Volunteer Fire Company. A group of men from the Piscatawaytown section of Raritan Township formed Raritan Engine Company No. 1 and the charter was signed on June 14, 1916, with 42 original members. The company provided fire protection for the southern portion of Raritan Township known as Fire District No. 1. The Menlo Park Hook and Ladder Fire Company formed in 1917 and, upon official recognition by the township seven years later changed its name to the Edison Volunteer Fire Company No. 1 in honor of Menlo Park's famous resident, Thomas Alva Edison. In 1923, Raritan Engine Company No. 2 was organized to provide fire protection in the Clara Barton section. Today Raritan Engine Company No. 2 is the only volunteer fire company in Edison that still operates out of its original firehouse. In 1927, two volunteer fire companies in the northern part of Raritan Township were formed: the H. K. Volunteer Fire Company and the Oak Tree Volunteer Fire Company.

Raritan Township was divided into five fire districts. Fire service for each district contained one of the previously mentioned volunteer fire companies and each district was managed by a board of five elected fire commissioners. Residents within a particular district paid a special fire tax that went towards operating expenses required by their local fire company. In 1926, the fire commissioners of District No. 1 hired a full-time firefighter to compensate for the lack of volunteer response during daytime hours. Throughout the years, more firefighters were hired, and by 1956, a total of 16 career firefighters were employed by Fire District No. 1 under the command of Chief Engineer Arthur Latham. The fire commissioners of District No. 3 were also inspired to hire career firefighters for the Clara Barton section of Raritan Township. As a result, two firefighters were hired in 1949 and three more were appointed within the next few years.

Like Edison Volunteer Fire Company No. 1, which had changed its name 30 years earlier, residents voted to change the township's name from Raritan to Edison in 1954. An even bigger change occurred in 1958 when the township changed from a commissioner-type of government to a mayor and council-type of government (under the Faulkner Act) consisting of a strong mayor and a weak council. This change in government had a direct impact on Edison's fire service. Fire districts and fire commissioners were abolished and the Edison Department of Public Safety was created containing three divisions: Police, Fire, and First Aid. The new and powerful mayor appointed a Division of Fire supervisor to manage and provide administrative needs to all five volunteer fire companies. At this time there were about 20 career firefighters between Fire Districts No. 1 and No. 2 who were now under the control of the Division of Fire supervisor. Within a short period of time, four career firefighters were promoted to the position of captain, which created the foundation for the hierarchical chain of command within the organization. Furthermore, terminology had changed from "districts" to "precincts," but geographical response areas remain the same. Raritan Engine Company No. 1 covers Precinct 1, Edison Volunteer Fire Company No. 1 covers Precinct 2, Raritan Engine Company No. 2 covers Precinct 3, H. K. Volunteer Fire Company covers Precinct 4, and Oak Tree Volunteer Fire Company covers Precinct 5.

Prior to the change in administration, the career firefighters from both districts decided to unite and form a union. They contacted the International Association of Fire Fighters (IAFF) and, after following proper procedures, received a charter in 1955. As a result, the career firefighters from both fire districts became collectively known as the Edison Township Fire Fighters Association, Local No. 1197.

In 1967, a fatal house fire occurred in northern Edison. Residents were distraught over the lengthy volunteer response time. Public officials held an open meeting at the Oak Tree School for residents to discuss the troubling situation and, as a result, career firefighters were staffed in North Edison. By the mid-1970s, all firehouses in Edison, except for Menlo Park, were staffed with career firefighters. In 1992, however, Edison Volunteer Fire Company No. 1 in Menlo Park could no longer provide adequate staffing, and career firefighters moved into the Menlo Park firehouse.

Those applying to become a career firefighter in the Edison Division of Fire are required to be an active Edison volunteer firefighter for a minimum of two years and to successfully complete a series of assessments, which include a written exam, a physical agility test, a civilian oral board interview, a fire-related oral board interview, a written essay, a medical physical examination, and a psychological examination. Edison Division of Fire career firefighters work rotating shifts that provide the township with 24-hour protection, seven days a week. In addition to the traditional duty of protecting lives and property from fire, today's Edison firefighters respond to medical emergencies, vehicle accidents, entrapments, elevator emergencies, confined space rescues, high angle rescues, structural collapses, and any other type of emergency that may occur.

In 2004, Mayor George Spadoro publicly announced plans of a new public safety facility (a firehouse/police substation) to be built in Raritan Center Industrial Park. The construction of the facility would incur no cost to taxpayers due to a special agreement in which the land, structure, and equipment will be donated by the two owners of Raritan Center, Summit Associates and Federal Business Center. This firehouse will become the seventh firehouse within the township.

The Edison Division of Fire has developed into one of the finest organizations in Middlesex County, New Jersey. Its members demonstrate their dedication and professionalism at every alarm and, therefore, I am truly honored to be a part of Edison's bravest.

One

Raritan Engine Company No. 1

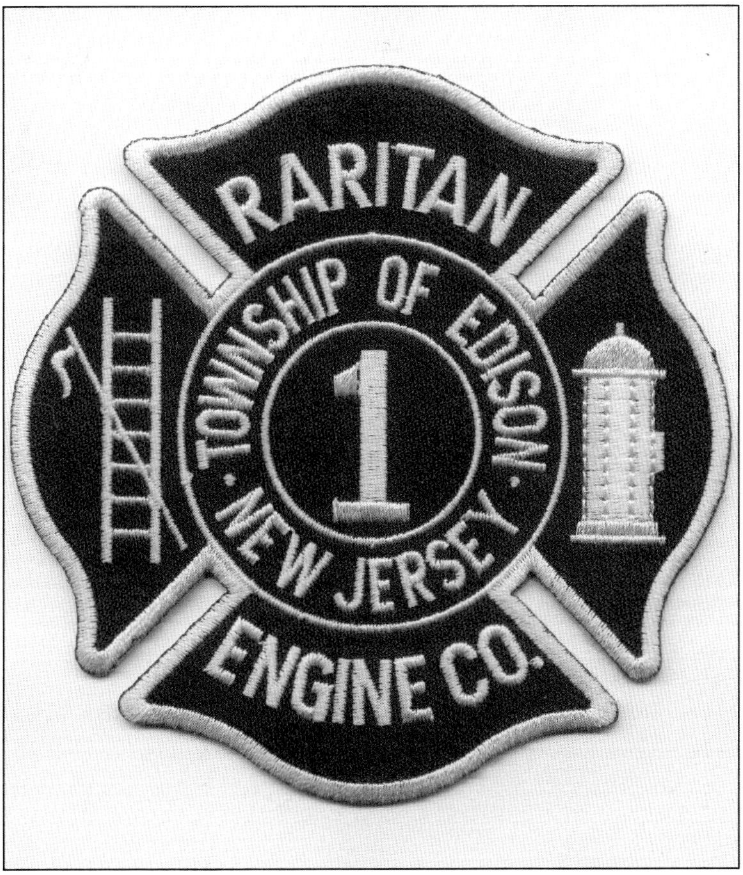

On June 17, 1916, Raritan Engine Company No. 1 received its charter and officially started providing fire protection to the Lindeneau, Piscatawaytown, Bonhamton, Nixon, and Stelton sections of Raritan Township. Today the company still protects these southern sections, which are currently known as Precinct 1.

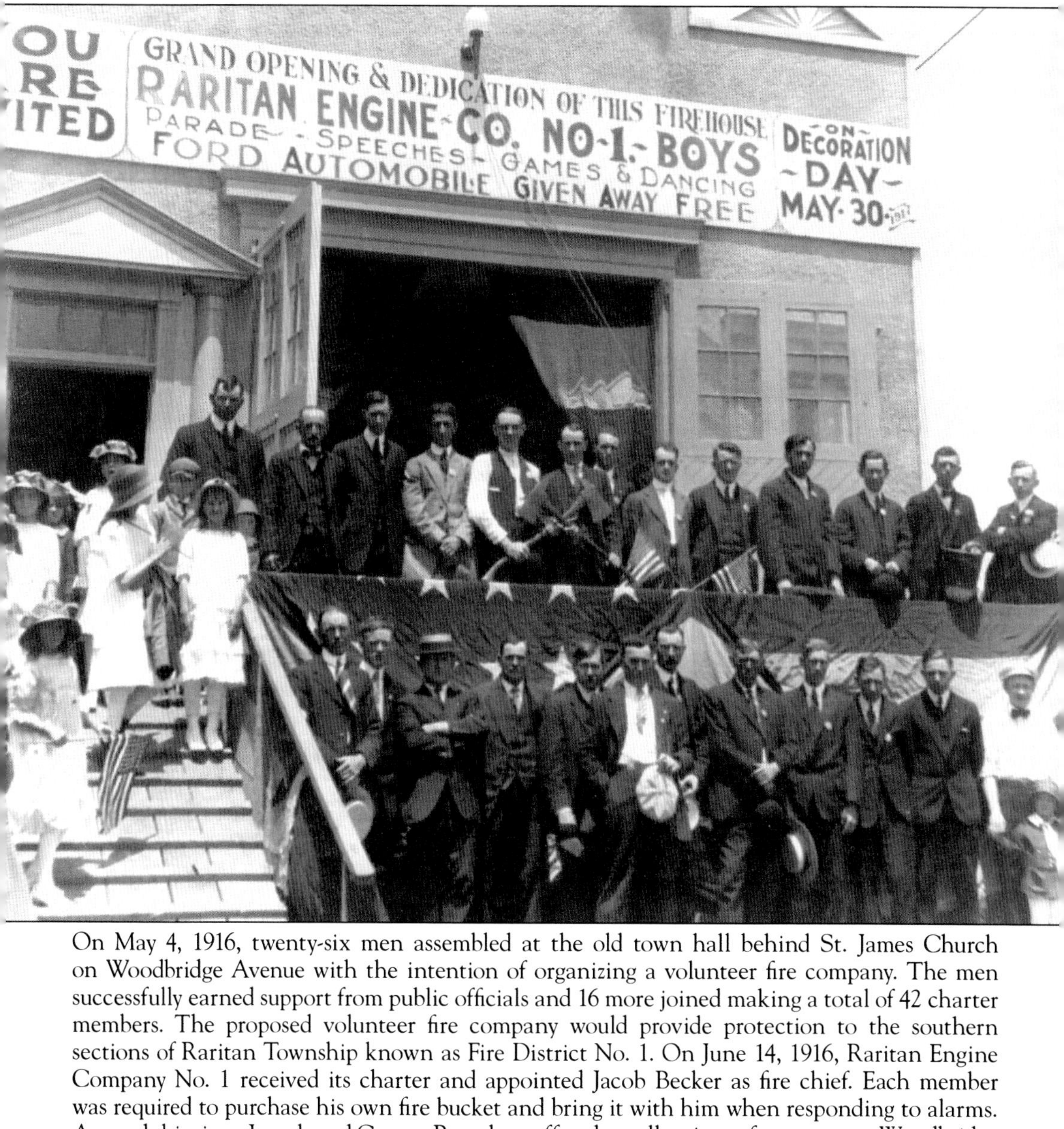

On May 4, 1916, twenty-six men assembled at the old town hall behind St. James Church on Woodbridge Avenue with the intention of organizing a volunteer fire company. The men successfully earned support from public officials and 16 more joined making a total of 42 charter members. The proposed volunteer fire company would provide protection to the southern sections of Raritan Township known as Fire District No. 1. On June 14, 1916, Raritan Engine Company No. 1 received its charter and appointed Jacob Becker as fire chief. Each member was required to purchase his own fire bucket and bring it with him when responding to alarms. Around this time, Joseph and George Brundage offered to sell a piece of property on Woodbridge Avenue for the sum of $200 to Raritan Engine Company No. 1. The company accepted the offer and by January 1917, a new firehouse was built for $2,880. Pictured above is the dedication of Raritan Engine Company No. 1's new firehouse on Decoration Day 1917. (Courtesy of Ruth Swales Bernat.)

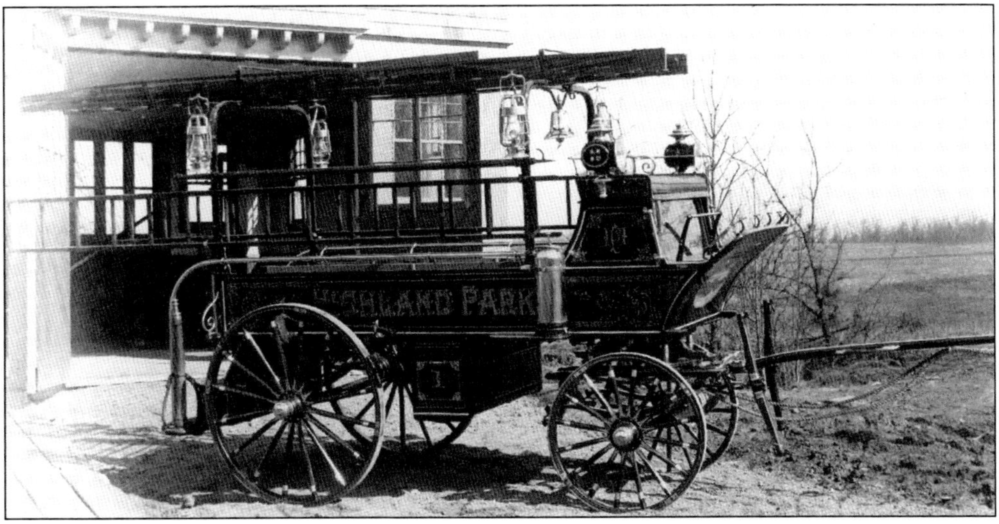

In 1917, the neighboring Highland Park Volunteer Fire Department helped the newly formed Raritan Engine Company No. 1 by lending its old horse-drawn hose wagon, shown above. The wagon not only carried hose, but was also equipped with nozzles, portable extinguishers, hand tools, and a wooden ladder. Within a short period of time, the commissioners of the newly formed Fire District No. 1 purchased a 1917 Ford Model T fire engine for the company. As soon as the company received the Ford, they returned the wagon to Highland Park. Pictured below are the original members of Raritan Engine Company No. 1 proudly posing with the borrowed hose wagon in front of their new firehouse on May 30, 1917. (Above, courtesy of Highland Park Volunteer Fire Department; below, courtesy of Ruth Swales Bernat.)

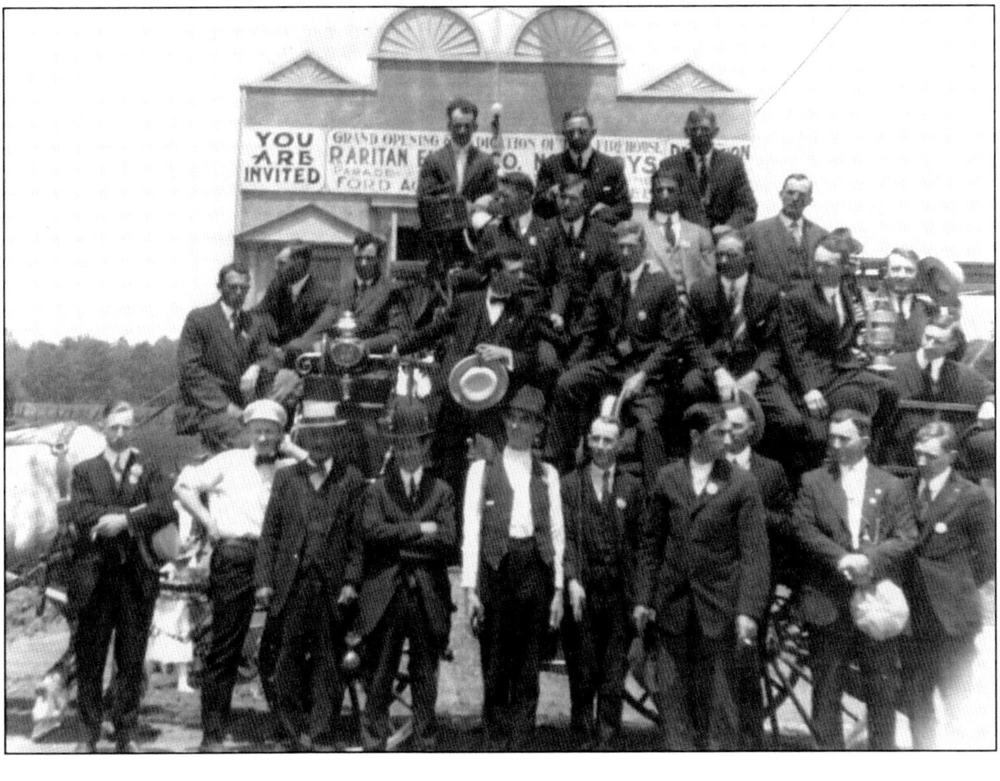

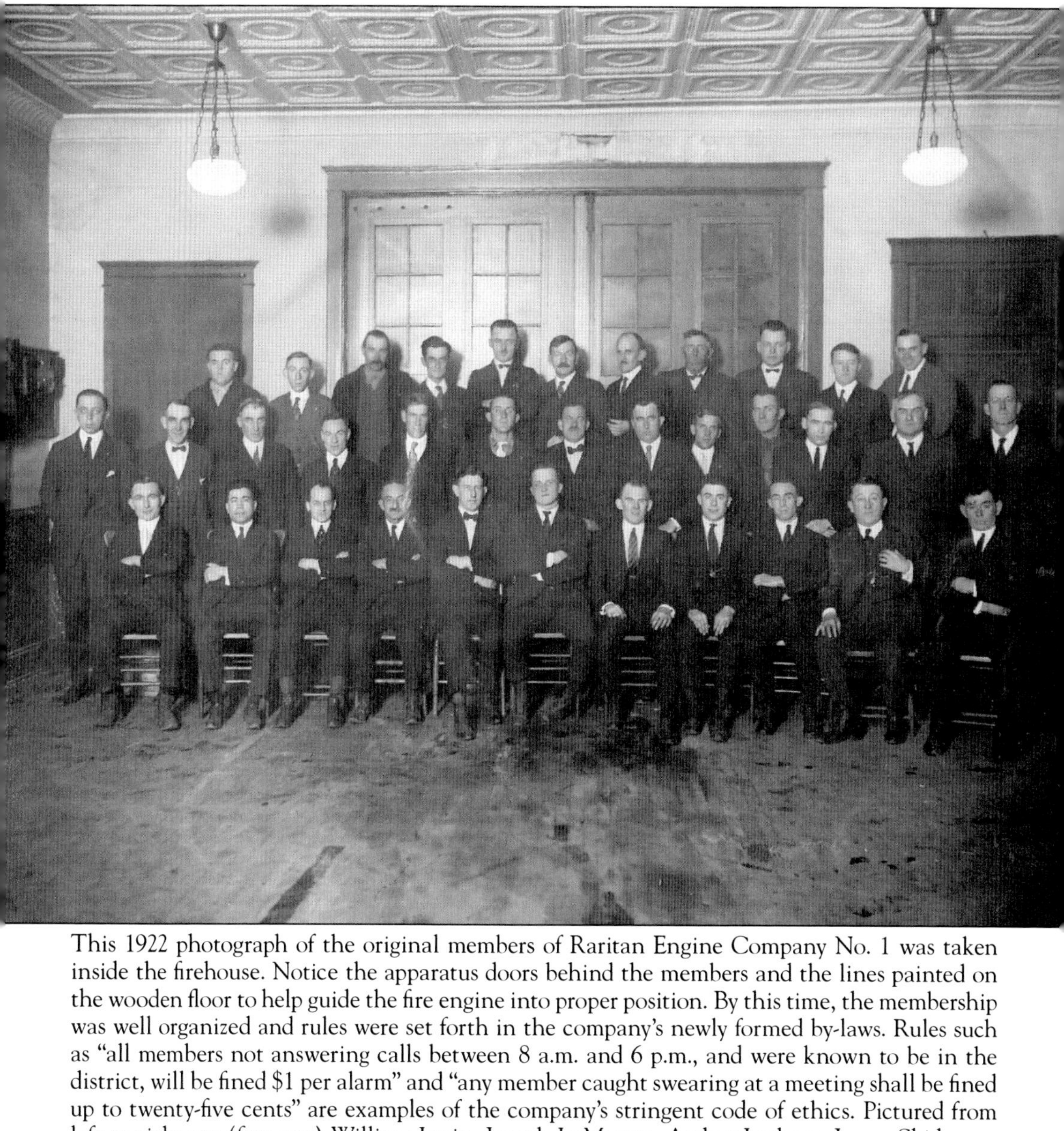

This 1922 photograph of the original members of Raritan Engine Company No. 1 was taken inside the firehouse. Notice the apparatus doors behind the members and the lines painted on the wooden floor to help guide the fire engine into proper position. By this time, the membership was well organized and rules were set forth in the company's newly formed by-laws. Rules such as "all members not answering calls between 8 a.m. and 6 p.m., and were known to be in the district, will be fined $1 per alarm" and "any member caught swearing at a meeting shall be fined up to twenty-five cents" are examples of the company's stringent code of ethics. Pictured from left to right are (first row) William Irwin, Joseph LoMastro, Arthur Latham, James Skidmore, Paul Berrue, Paul Fisher, Joseph Pettit, James O'Connor, Herbert Wildgoose, Alexander Bors, and Nels Nelson; (second row) William Kuehl, Charles Fredricks, Thomas Ward, Leo Fochman, George Rush, Fred Schulteiss, Peter Meeker, Manning Drake, James Stalker, Eugene McDonald, Bernard O'Hara, John Perdun, and Frank White; (third row) Frank Sansouci, Max Fochman, Francis Woerner, Frederick Dumas, Edward Starkin, Wilfred R. Woodward, Fred Pettit, Van Stout, Remsen Hansmann, Edward Voorhees, and Joseph Fetzer. (Courtesy of Raritan Engine Company No. 1.)

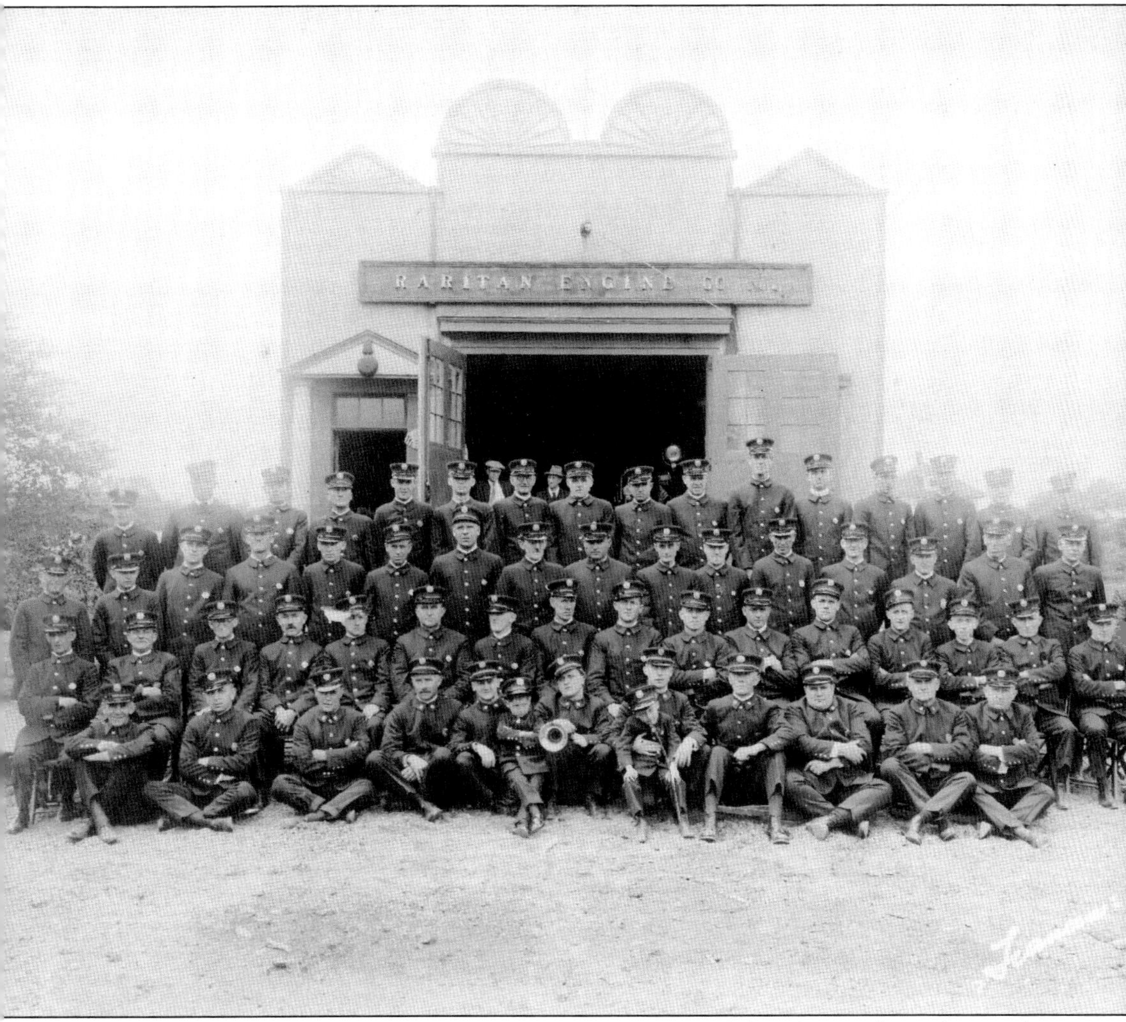

In 1925, members of Raritan Engine Company No. 1 purchased their first uniforms from Cohen and Company of Newark for $29 each. Since the company's general fund was limited, each member was required to pay $10 toward the acquisition of their uniform. Posing in front of the firehouse wearing their new uniforms are (first row) Charles Fredericks, William Kuehl, Thomas Swales, Francis Woerner, Arthur Latham, Arthur Latham Jr., Paul Fisher, young Donald Rockhill sitting on James O'Connor's lap, Eugene McDonald, Philip Beaudoin, John Perdun, and Alexander Bors; (second row) Walter DeKyne, Peter Meeker, William Messerole, Robert Rohr, William Enoch, Manning Drake, Joseph Pettit, George Rush, Frederick Schultheiss, Arthur Nemis, Charles Pfeiffer, Russell Rockhill, William Fisher, Max Fochman, William Irwin, and Andrew Kane; (third row) Stanley Van Sickle, Arthur Perdun, Matthew Simpson, Frank White, Bernard O'Hara, Clarence Stout, Leo Myers, Frederick Pettit, Herbert Pfeiffer, Frederick Dumas, Leo Fochman, Thomas Ward, Charles Folger, Edward Voorhees, John Van Stout, Remsen Hansmann, and Albert Davis. The fourth row includes Albert Fredericks, Julius Engel, Nels Nelson, Charles Benni, Louis Pettit, Christopher Doll, Frank Sansouci, Anthony Governale, Joseph Fetzer, Andrew Bedle, Frederick Newman, Alexander Hazara, and Arnold Niehaus. (Courtesy of Duane Clause.)

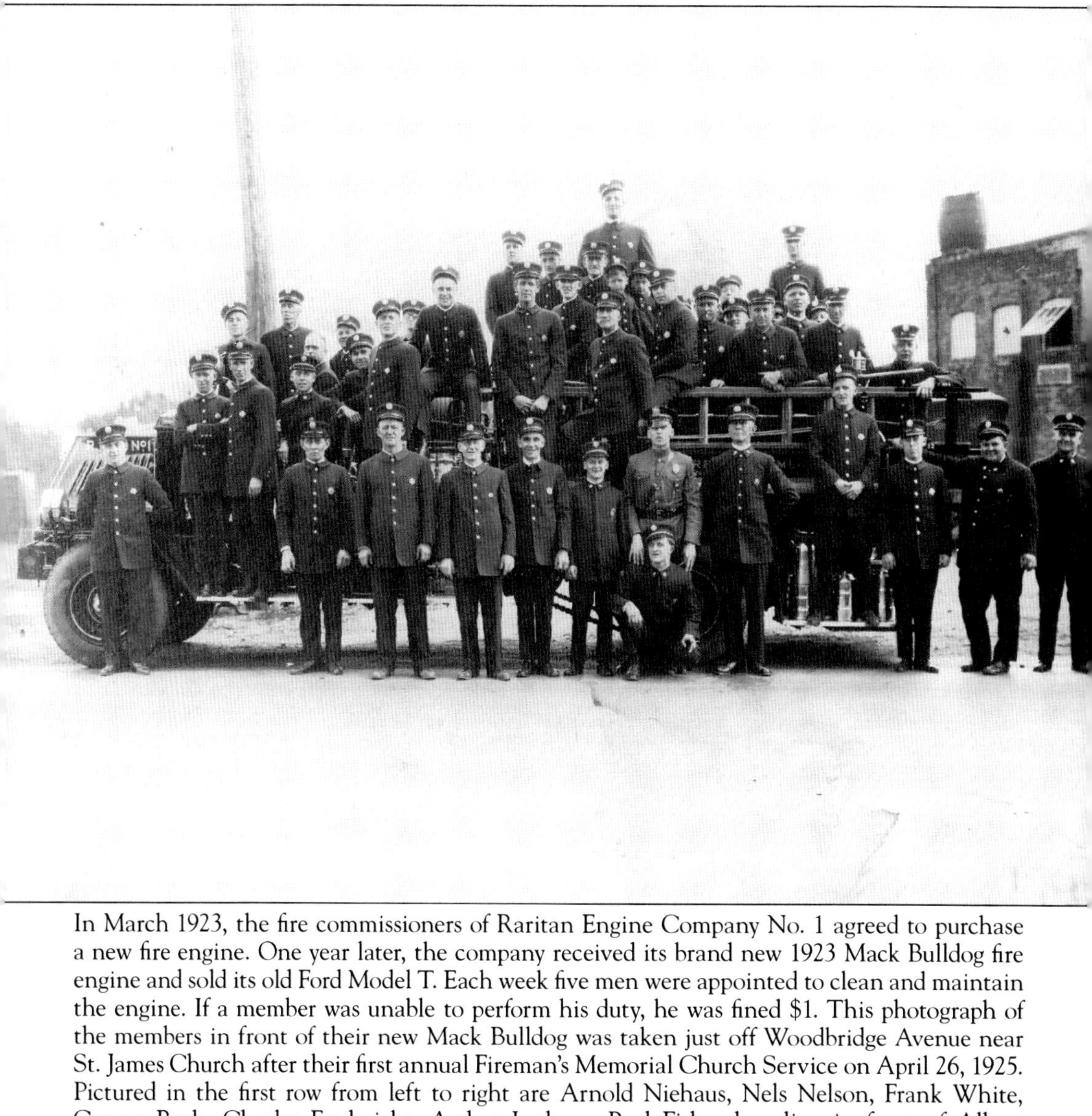

In March 1923, the fire commissioners of Raritan Engine Company No. 1 agreed to purchase a new fire engine. One year later, the company received its brand new 1923 Mack Bulldog fire engine and sold its old Ford Model T. Each week five men were appointed to clean and maintain the engine. If a member was unable to perform his duty, he was fined $1. This photograph of the members in front of their new Mack Bulldog was taken just off Woodbridge Avenue near St. James Church after their first annual Fireman's Memorial Church Service on April 26, 1925. Pictured in the first row from left to right are Arnold Niehaus, Nels Nelson, Frank White, George Rush, Charles Fredericks, Arthur Latham, Paul Fisher kneeling in front of Albert Davis, Eugene McDonald, William Fisher, Joseph Stout, Philip Beaudoin, and John Perdun. The members behind them include Clarence Stout, George Rush, James O'Connors, Frederick Schulteiss, Andrew Bedle, Charles Benni, Louis Pettit, Arthur Nemis, Van Stout, Peter Meeker, Anthony Governale, Chris Doll, Russell Rockhill, Frederic Dumas, Joseph Pettit, Julius Engel, William Keuhl, Herbert Pfeiffer, Leo Fochman, Manning Drake, Leo Meyers, Thomas Ward, Larry Guadagnino, Rem Hansmann, and Andrew Kane. (Courtesy of Raritan Engine Company No. 1.)

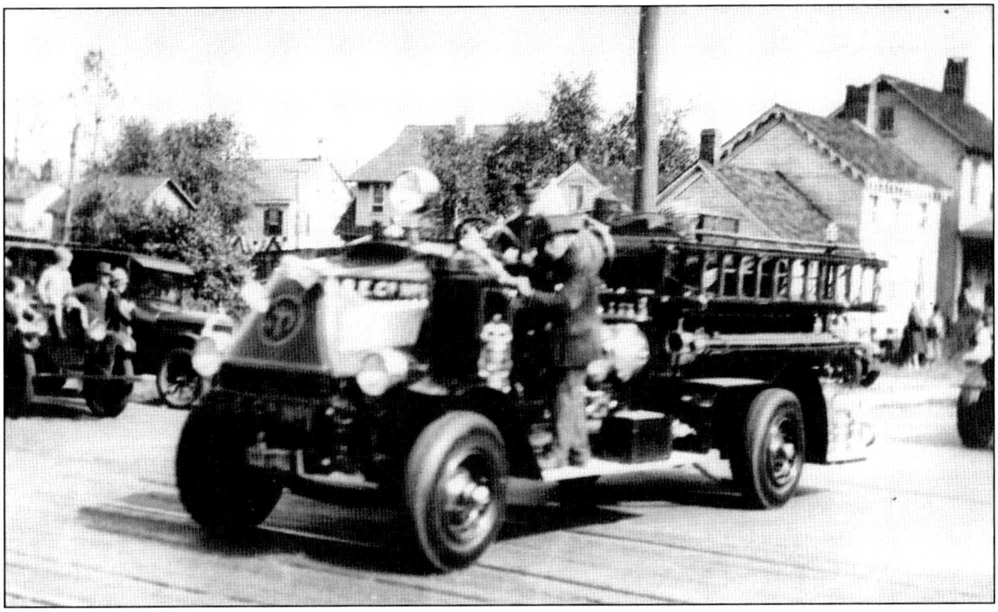

This photograph, which dates back to the mid-1920s, shows the Mack Bulldog fire engine driving across an unfamiliar bridge. The Mack remained in service until 1954 when the fire commissioners decided to sell it. Fellow member Thomas Swales Jr. purchased the apparatus and sold it back to the company for $1 with the stipulation that it would remain as an integral part of Raritan Engine Company No. 1. (Courtesy of Duane Clause.)

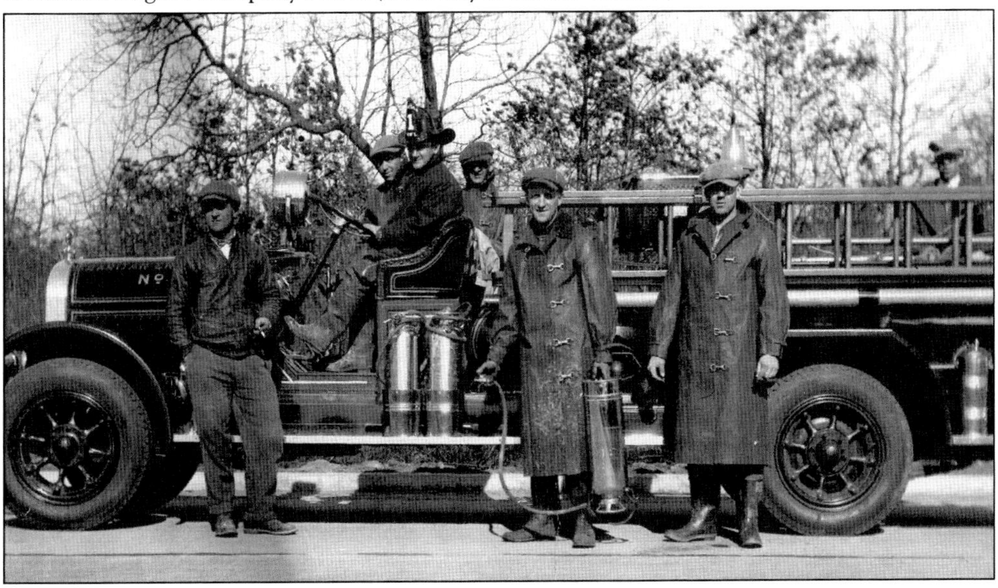

In 1925, Raritan Engine Company No. 1 purchased a pre-owned Cosmopolitan fire engine to extinguish field fires and, therefore, conserve the Mack Bulldog fire engine for building fires and parades. The members kept the Cosmopolitan engine, which they referred to as the "Old Dinky," in service until 1937, when a new Ford was purchased. In the picture above, standing in front from left to right are Paul Fisher, William Fisher Sr., and James Kane; behind the wheel is Arthur Latham and in the passenger seat is Clarence Stout; standing in the back is Joseph Horvath. (Courtesy of Duane Clause.)

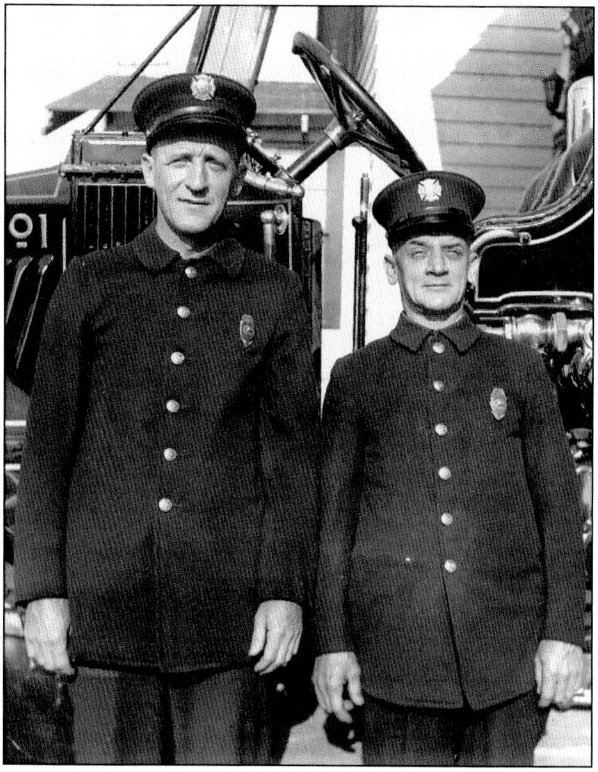

Chief Engineer Arthur W. Latham (right) and Assistant Chief Engineer William Fisher (left) pose in front of the Mack fire engine. Latham, who was hired in 1926 by the commissioners of Fire District No. 1, was the first career firefighter of Raritan Township (now known as Edison). Fisher was hired two years later as Latham's assistant and retired after 25 years of service in 1953. (Courtesy of Brian Latham.)

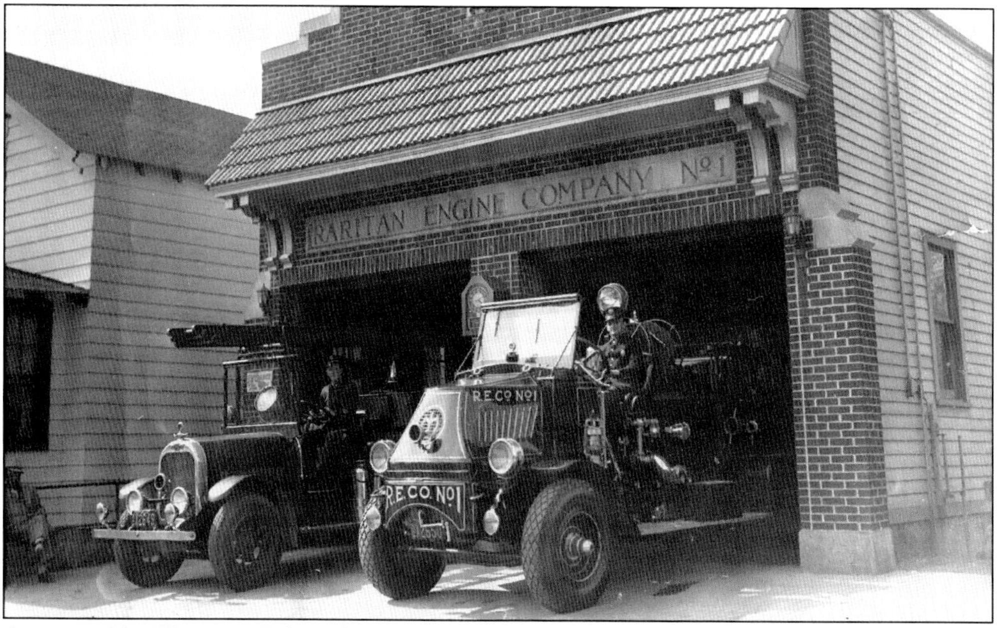

Assistant Chief Engineer Fisher is behind the wheel of the Cosmopolitan (left) and Chief Engineer Latham is behind the wheel of the 1923 Mack Bulldog (right). This picture was taken after renovations were made to the original Raritan Engine Company No. 1 firehouse on Woodbridge Avenue. In addition to the new brick front and double doors, a concrete floor was poured because the new Mack was too heavy for the wooden floor. (Courtesy of Brian Latham.)

Ruth Swales, standing with her arms around her sons Thomas (left) and James, is the only woman to be voted as an honorary member of Raritan Engine Company No. 1. After building their house directly next door to the firehouse on Woodbridge Avenue, the Swales family installed a separate telephone line dedicated to emergency calls. Ruth Swales would receive the emergency telephone calls from inside her house and then go out to strike the iron gong in front of the firehouse in order to alert the firefighters. (Courtesy of Ruth Swales Bernat.)

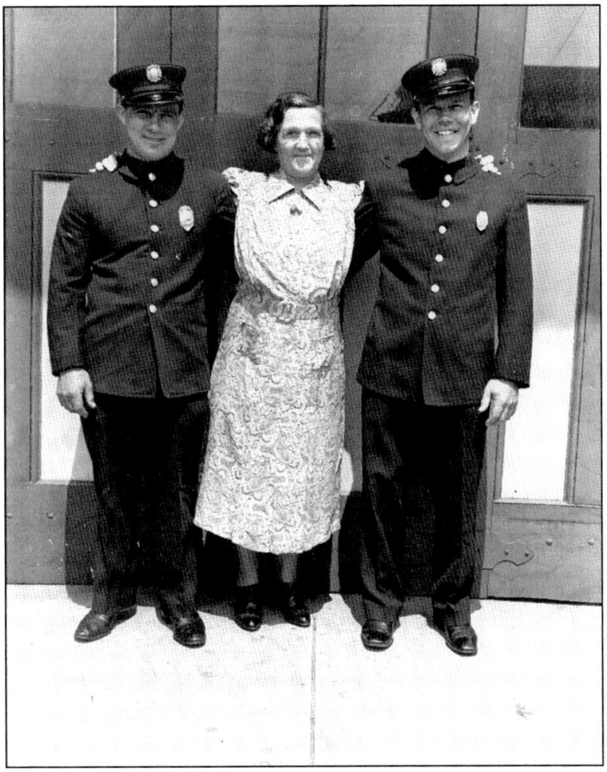

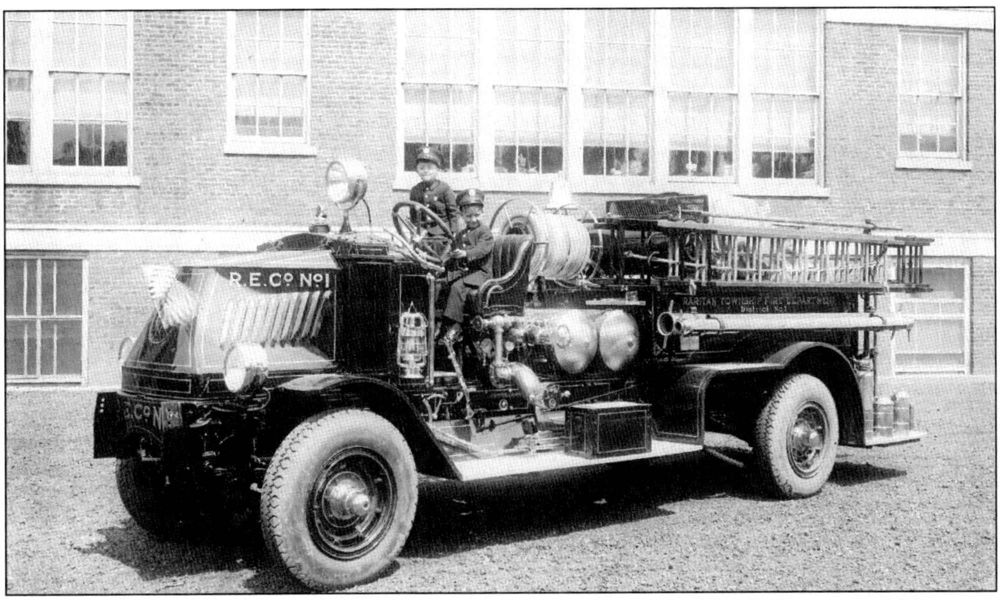

This 1925 photograph was taken after Raritan Engine Company No. 1's first annual Fireman's Memorial Church Service, which was held at the St. James Church on Woodbridge Avenue. The membership decided there would be a special church service on the last Sunday in April devoted to honoring their deceased members. Pictured above are youngsters Arthur Latham Jr. (left) and Donald Rockhill ready for action while sitting in the driver's seat of the Mack Bulldog alongside Piscatawaytown School No. 3. (Courtesy of Raritan Engine Company No. 1.)

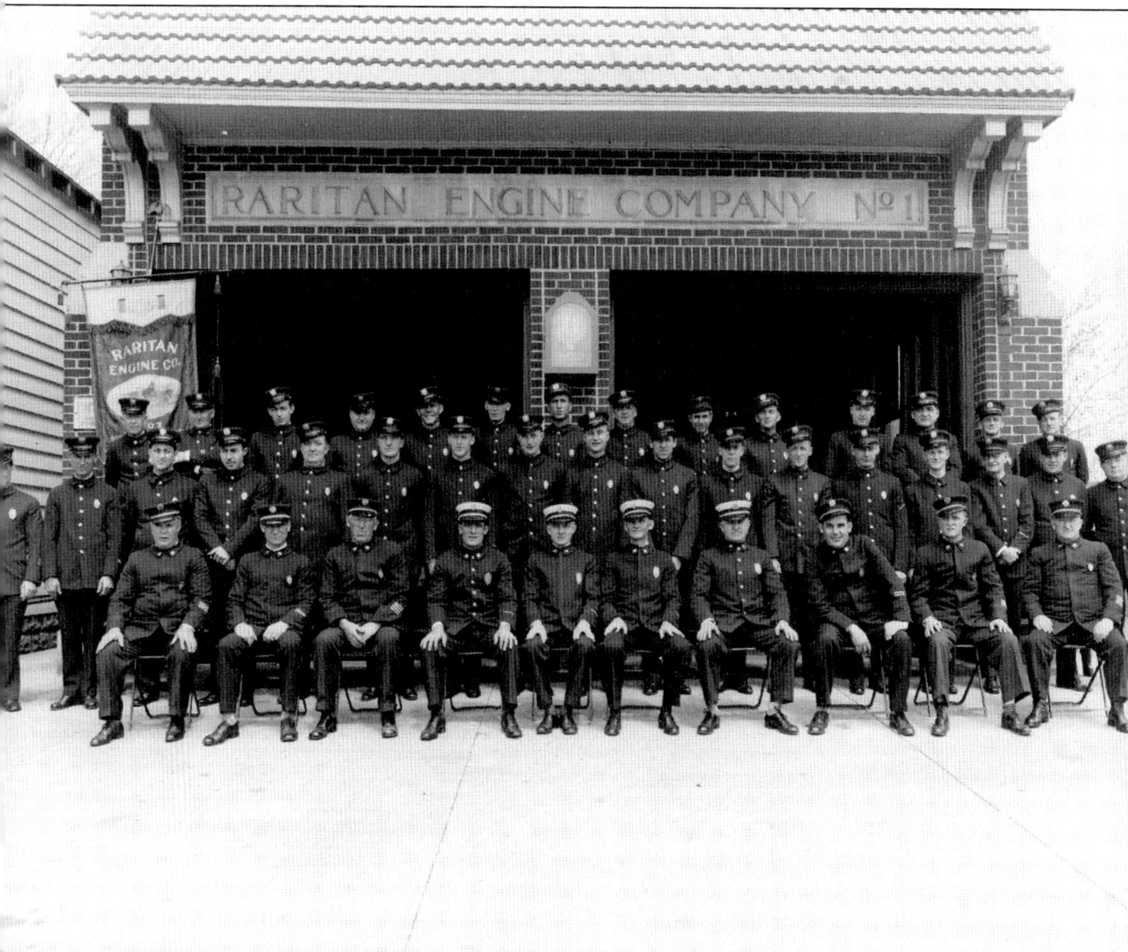

After attending their annual Fireman's Memorial Church Service, the members of Raritan Engine Company No. 1 pose in front of their firehouse on April 26, 1936. This was the last group photograph taken in front of the original firehouse on Woodbridge Avenue and the last group photograph wearing the old-style Class-A uniforms. Pictured from left to right are (first row) Thomas Swales Sr., Charles Folger, John Van Stout, Charles Oliveri, Chief Wendell Slavick, Theodore Eggertson, George Graff, Eiler Rasmussen, Paul Fisher, and Edward Voorhees; (second row) Thomas Swales Jr., Alexander Hazara, John Paladino, Oscar Pillar, Julius Engel, Frank Gavin, Peter Bochman, William Doll, Robert Ellmyer, Frank Takacs, Ezra Grant, William Fisher, William Fercho, Kenneth Rush, Christopher Doll, Joseph Costa, and Arnold Kneehouse; (third row) Albert Fredericks, Jack Powers, Eldon Rush, Stanley Van Sickle, Thomas Swales Jr., Bernard O'Hara, unidentified, Edward Monaghan, Anthony Governale, unidentified, James Monaghan, Joseph Ambrosio, Arthur Latham, and William I. Enoch. (Courtesy of Raritan Engine Company No. 1.)

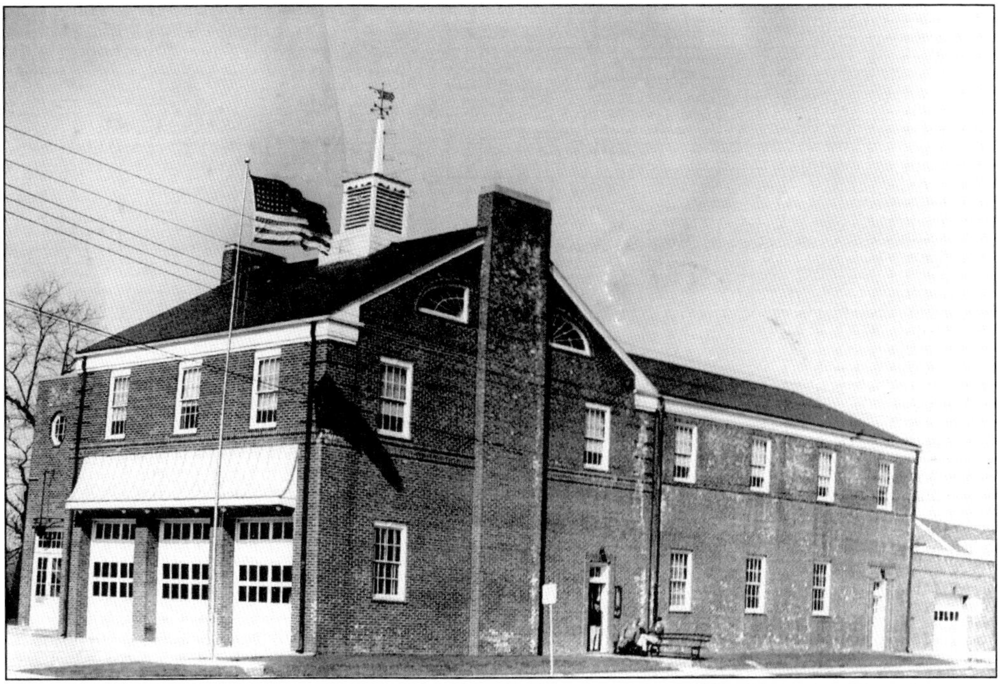

In 1938, the Board of Commissioners of Fire District No. 1 purchased land from Matthew Simpson for the sum of $2,000 to build a new firehouse. They acquired federal funding from the Works Progress Administration (WPA) to assist with the construction. Shown above, on the corner of Plainfield Avenue and Simpson Terrace, is the newly erected firehouse, which was photographed by local Piscatawaytown resident Remson Kentos. (Courtesy of Raritan Engine Company No. 1.)

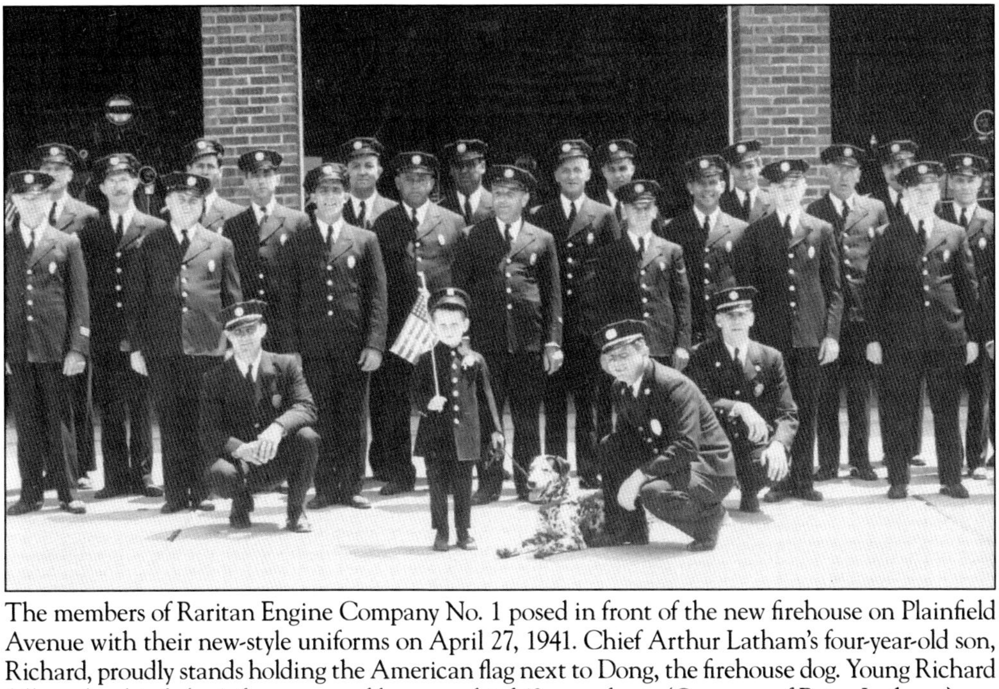

The members of Raritan Engine Company No. 1 posed in front of the new firehouse on Plainfield Avenue with their new-style uniforms on April 27, 1941. Chief Arthur Latham's four-year-old son, Richard, proudly stands holding the American flag next to Dong, the firehouse dog. Young Richard followed in his father's footsteps and became chief 42 years later. (Courtesy of Brian Latham.)

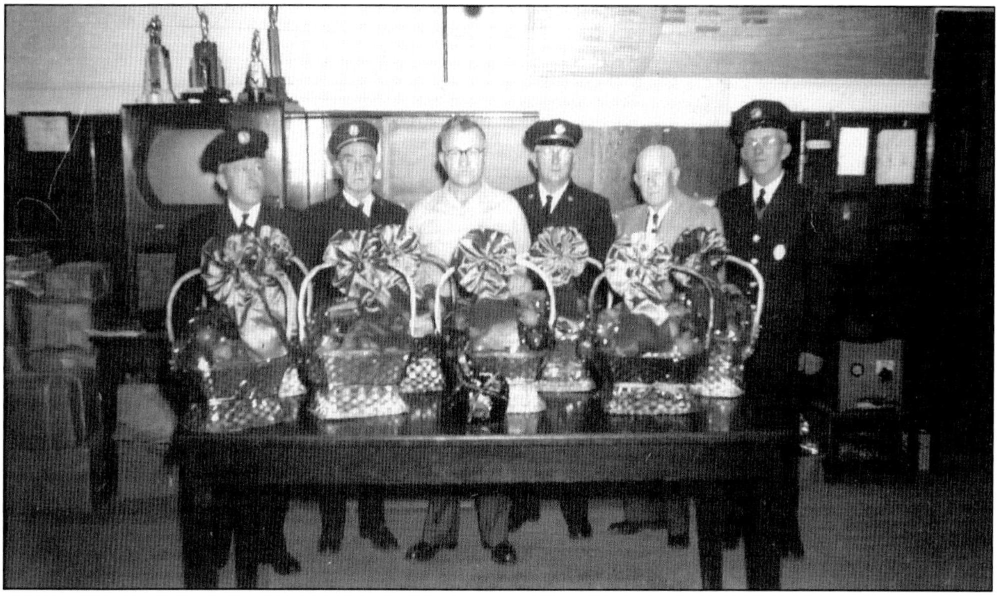

Throughout the 1940s and 1950s, Raritan Engine Company No. 1 prepared fruit baskets for the "shut-ins" during the holiday season. The term *shut-ins* referred to the sick and elderly members of the company. Pictured from left to right are Edward Voorhees, Herbert Wildgoose, George Graff, chaplain Jack Powers, Albert Fredericks, and Clarence Lowe. (Courtesy of Raritan Engine Company No. 1.)

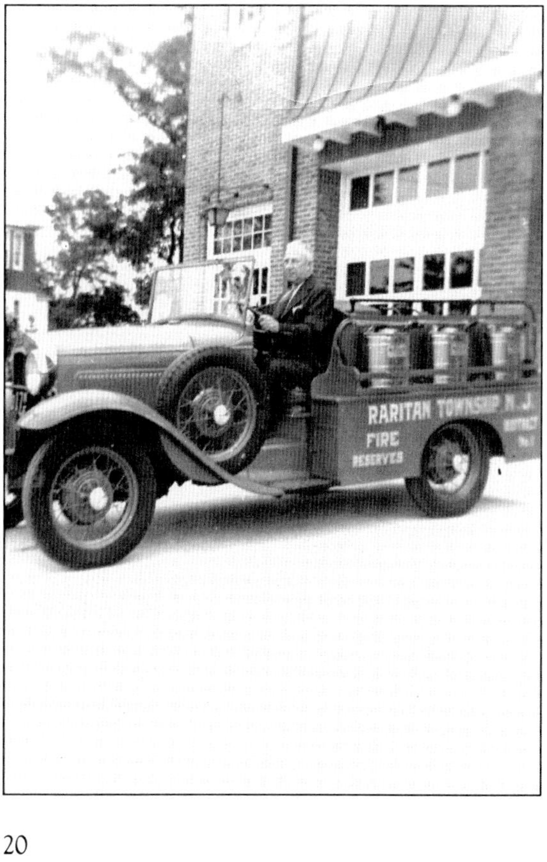

During the 1940s, civilian volunteers, known as the Raritan Township Fire Reserves, assisted with firefighting duties while most of the firefighting force left to fight in World War II. This photograph shows a member of the Fire Reserves, John Ellmeyer, sitting in a modified Ford Model A equipped with Indian Tanks. An Indian Tank is a portable water tank, held on a firefighter's back by shoulder straps, which was used to extinguish brush fires. Ellmeyer's son, John Jr., became Edison's chief of police in 1959.

The dalmatian, the breed associated with the fire service, was found in almost every firehouse during the early days of horse-drawn fire apparatus. These dogs were used not only to guide the horses, but also to protect the firehouse. During the 1930s and 1940s, Raritan Engine Company No. 1 had two dalmatians: Ding and Dong. This photograph shows one of the loyal canines in front of the fire reserve vehicle. (Courtesy of Duane Clause.)

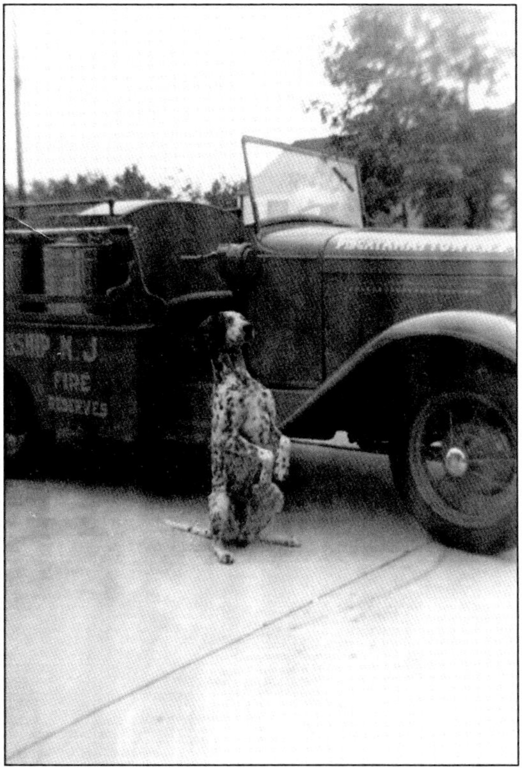

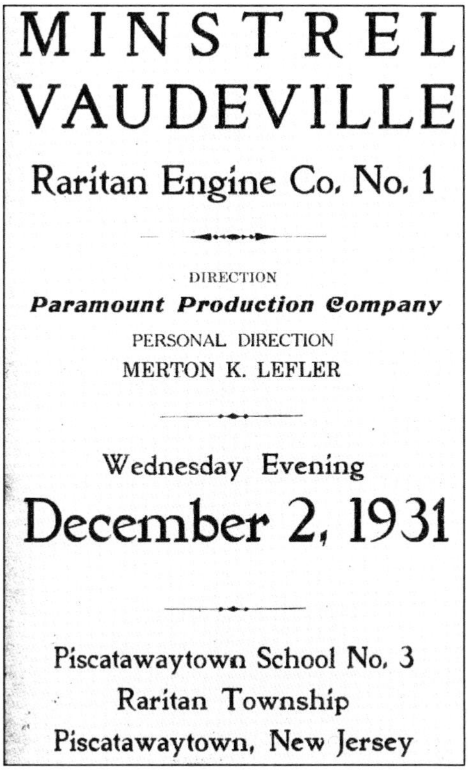

MINSTREL
VAUDEVILLE

Raritan Engine Co. No. 1

DIRECTION

Paramount Production Company

PERSONAL DIRECTION

MERTON K. LEFLER

Wednesday Evening

December 2, 1931

Piscatawaytown School No. 3
Raritan Township
Piscatawaytown, New Jersey

Throughout the years, it was common for fire companies to hold annual minstrel shows in order to raise money for equipment and operating expenses. The shows consisted of local firefighters performing amateur skits choreographed with music and dance. Raritan Engine Company No. 1 held their first minstrel on December 2, 1931, and continued the tradition until 1959. (Courtesy of Robert and Janet Grand-Jean.)

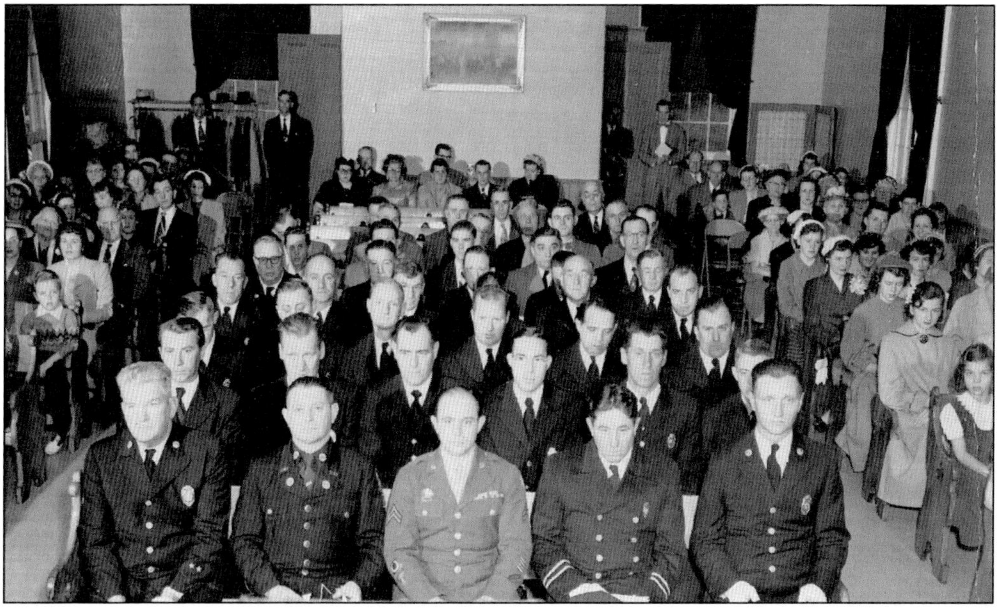

In the late 1940s, the members of Raritan Engine Company No. 1 decided to change the date of their annual Fireman's Memorial Church Service from the last Sunday in April to the first Sunday in May, and to rotate the location of the service between the churches located within their district. Shown above are members gathered together inside the Piscatawaytown Baptist Church, which is located on the corner of Woodbridge Avenue and Chapel Street, on May 3, 1953. (Courtesy of H. Ray Vliet.)

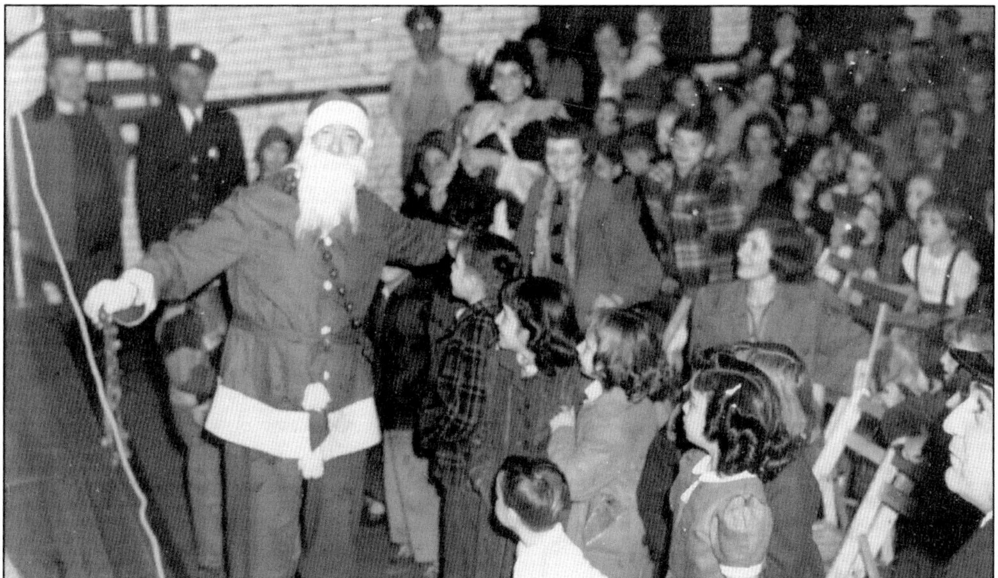

In its early days, Raritan Engine Company No. 1 hosted annual Christmas parties for children in its district that were up to 12 years of age. Unfortunately, this tradition was discontinued in the mid-1950s when the population grew to an unmanageable amount. Pictured above is Santa Claus (Julius Engel) greeting the children at an early-1950s Christmas party inside Piscatawaytown School No. 3. (Courtesy of Raritan Engine Company No. 1.)

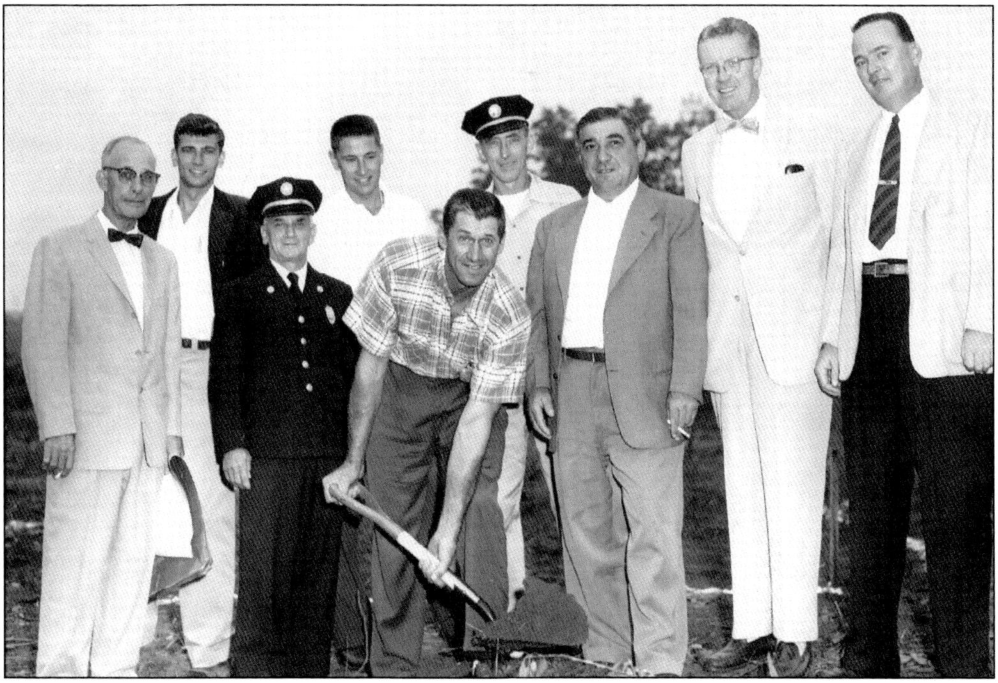

Like many regions throughout the nation, Fire District No. 1 experienced a rapid growth in population after World War II. New residential neighborhoods were quickly being developed in areas that had been desolate. In 1955, the fire commissioners asked Raritan Engine Company No. 1 to open its charter in order to increase its membership roster. The commissioners also searched for a plot of land in order to build another firehouse for the district. They acquired a lot on Lincoln Highway, which was northwest of their firehouse on Plainfield Avenue (Station No. 1) and geographically appropriate. Pictured above is the groundbreaking ceremony of Station No. 2. From left to right are A. Rasmussen (architect), A. Garlatti (contractor), Chief A. Latham, J. Garlatti (contractor), F. Takas (president), Assistant Chief H. Drake, J Ambrosio (vice president), A. Lawlor Jr. (treasurer), and B. Conerty (secretary). Shown below is a photograph of Station No. 2 taken during the 1980s. (Above, courtesy of Brian Latham; below, courtesy of Scott McCloughan.)

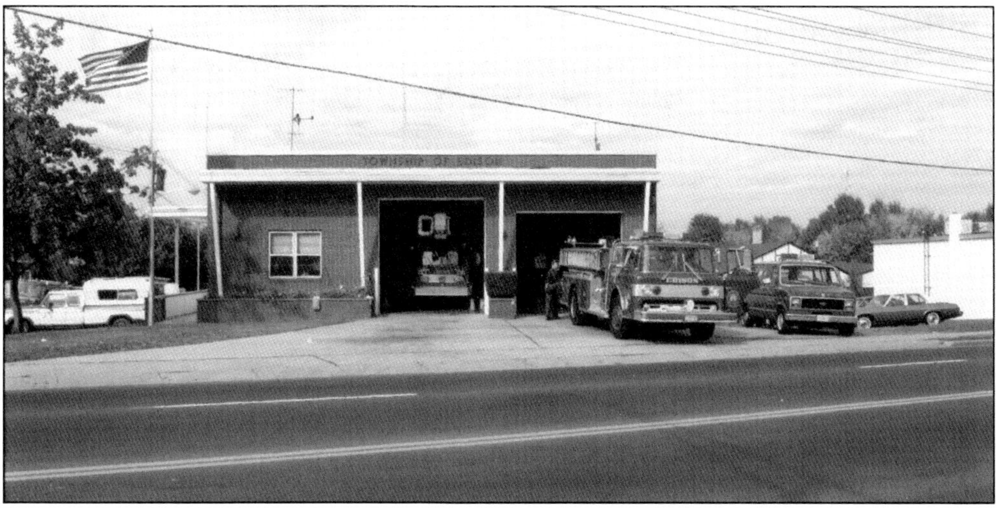

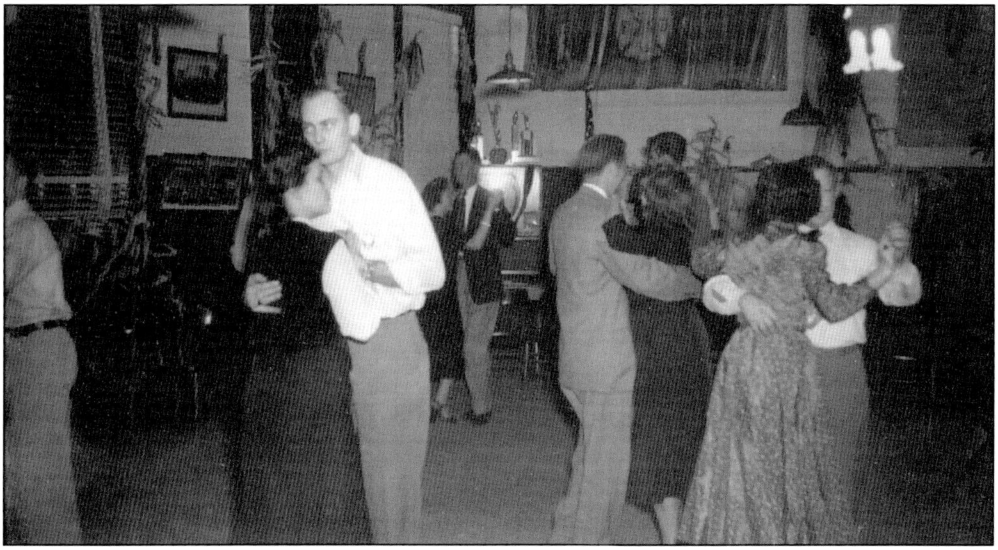

Firehouses have often been the center of many festive occasions. Members of Raritan Engine Company No. 1 and their family, friends, and community celebrated Christmas and New Year's Eve parties, anniversaries, birthdays, and many other special events at the firehouse. This 1951 photograph, which was taken on the second floor of Station No. 1, depicts a typical New Year's Eve celebration involving a lot of dancing. (Courtesy of Raritan Engine Company No. 1.)

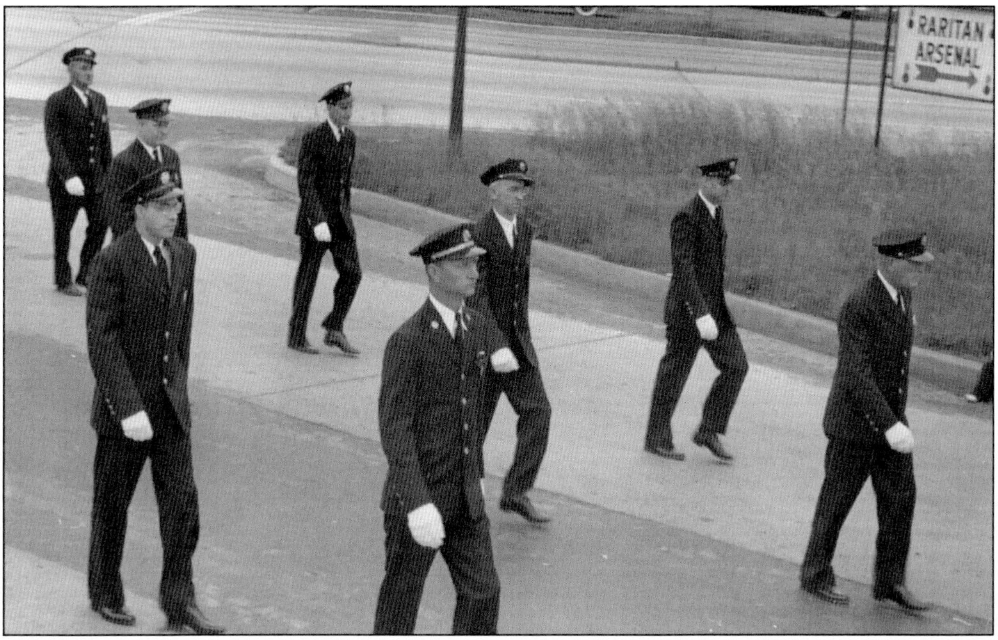

In this photograph, members of Raritan Engine Company No. 1 are marching up Woodbridge Avenue just past the intersection of Route 1 during the 1959 Memorial Day parade. Pictured from left to right are Anthony Cherenini, Kenneth Graff, James Burke, George Ellmeyer, Robert Ellmeyer, Clarence Lowe, John Nemeth, and Robert Bishop. Notice the Raritan Arsenal sign pointing up Woodbridge Avenue. The Raritan Arsenal was a large military base that was active up until the early 1960s. The base has been replaced by Raritan Center Industrial Park and Middlesex County College. (Courtesy of Raritan Engine Company No. 1.)

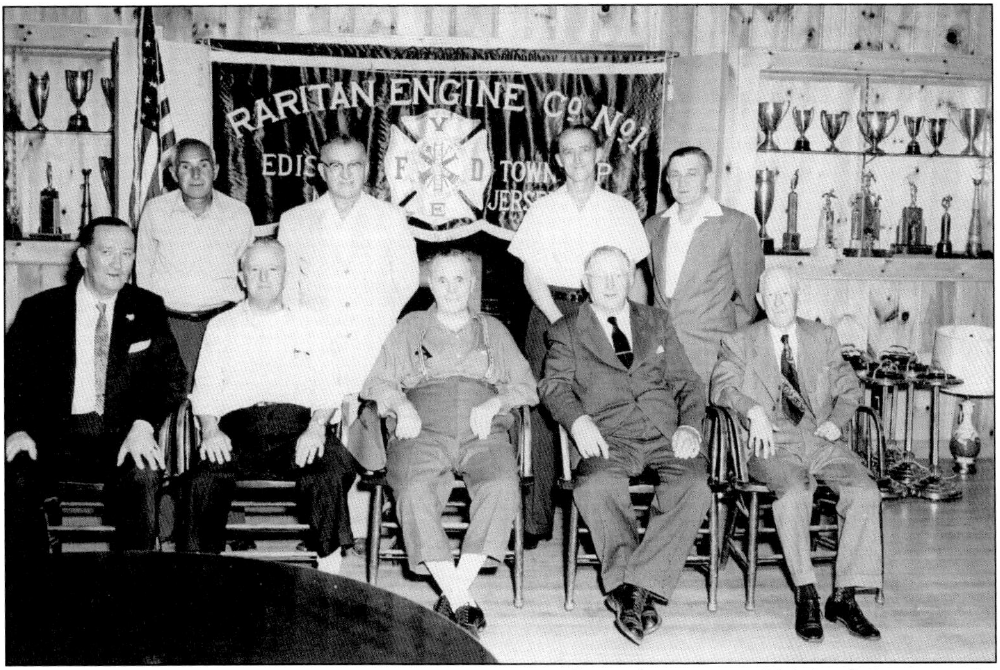

In 1959, members of Raritan Engine Company No. 1 hosted a surprise 75th birthday party for Thomas "Pop" Swales. Pop Swales, along with the rest of his family, has helped the company in numerous ways throughout the years. Pictured from left to right are (first row) Julius Engel, Edward Voorhees, Pop Swales, chaplain Jack Powers, and Albert Fredericks; (second row) Anthony Governale, George Graff, Harold Drake, and Wendell Slavick. (Courtesy of Raritan Engine Company No. 1.)

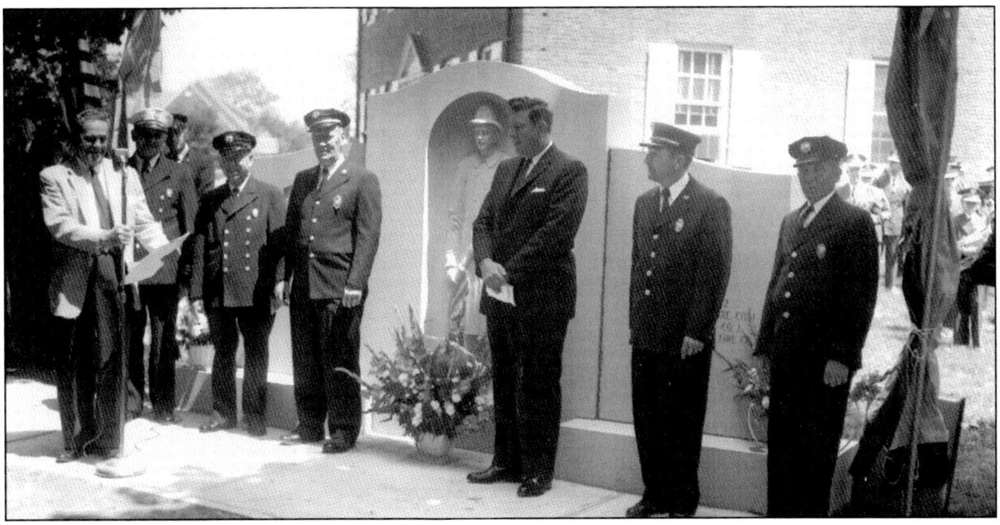

This 1961 photograph shows the dedication of the Edison Exempts Association monument. The monument was built alongside town hall near Simpson Avenue to honor all exempt firefighters of Edison Township. All township firefighters who attend a minimum of 60 percent of fires and drills for seven years are eligible for an exempt certification, which had originally provided three main benefits: exemption from jury duty, tenure of office, and a free hawker's license. (Courtesy of Raritan Engine Company No. 1.)

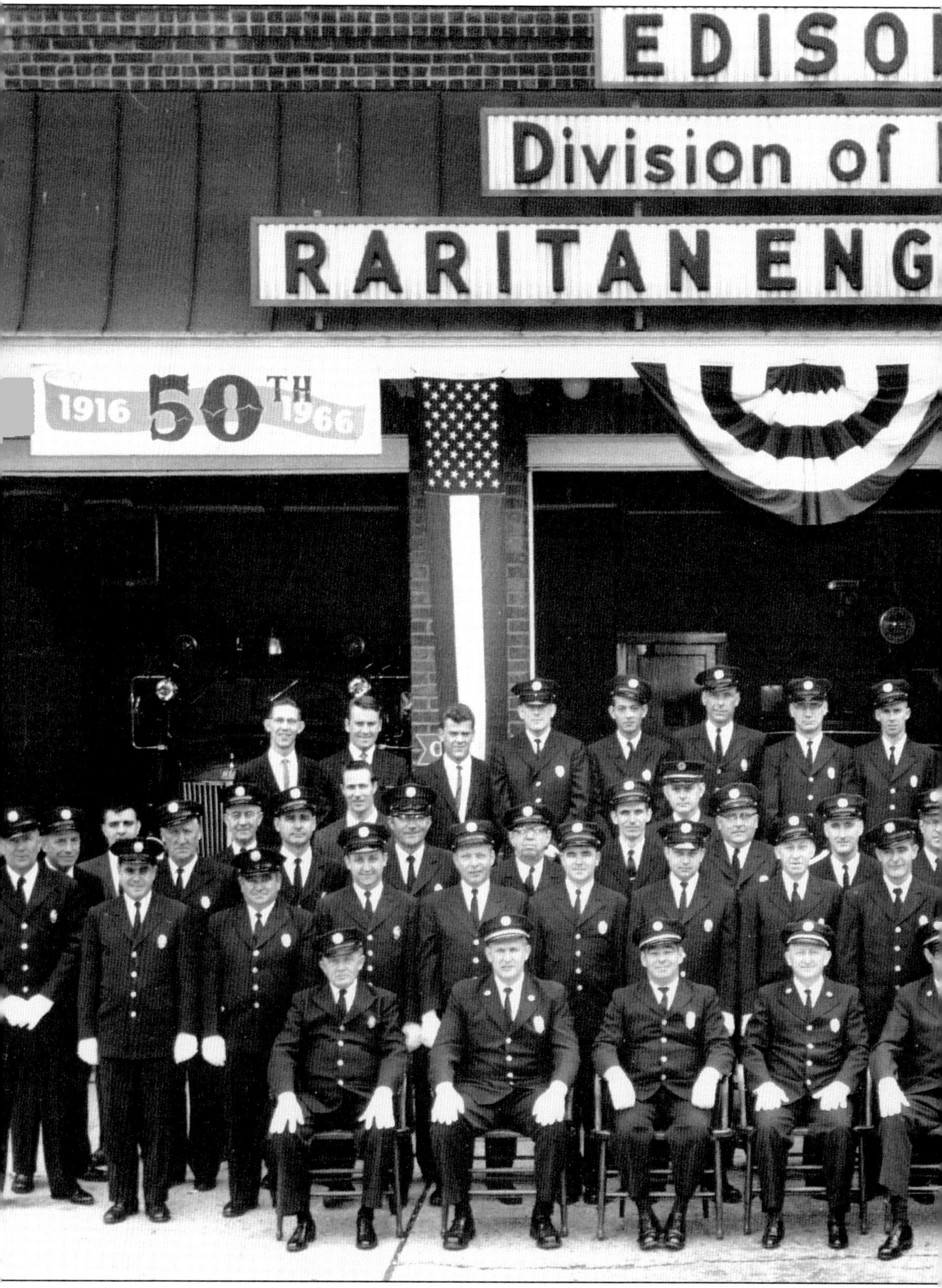
EDISO
Division of
RARITAN ENG
1916 50TH 1966

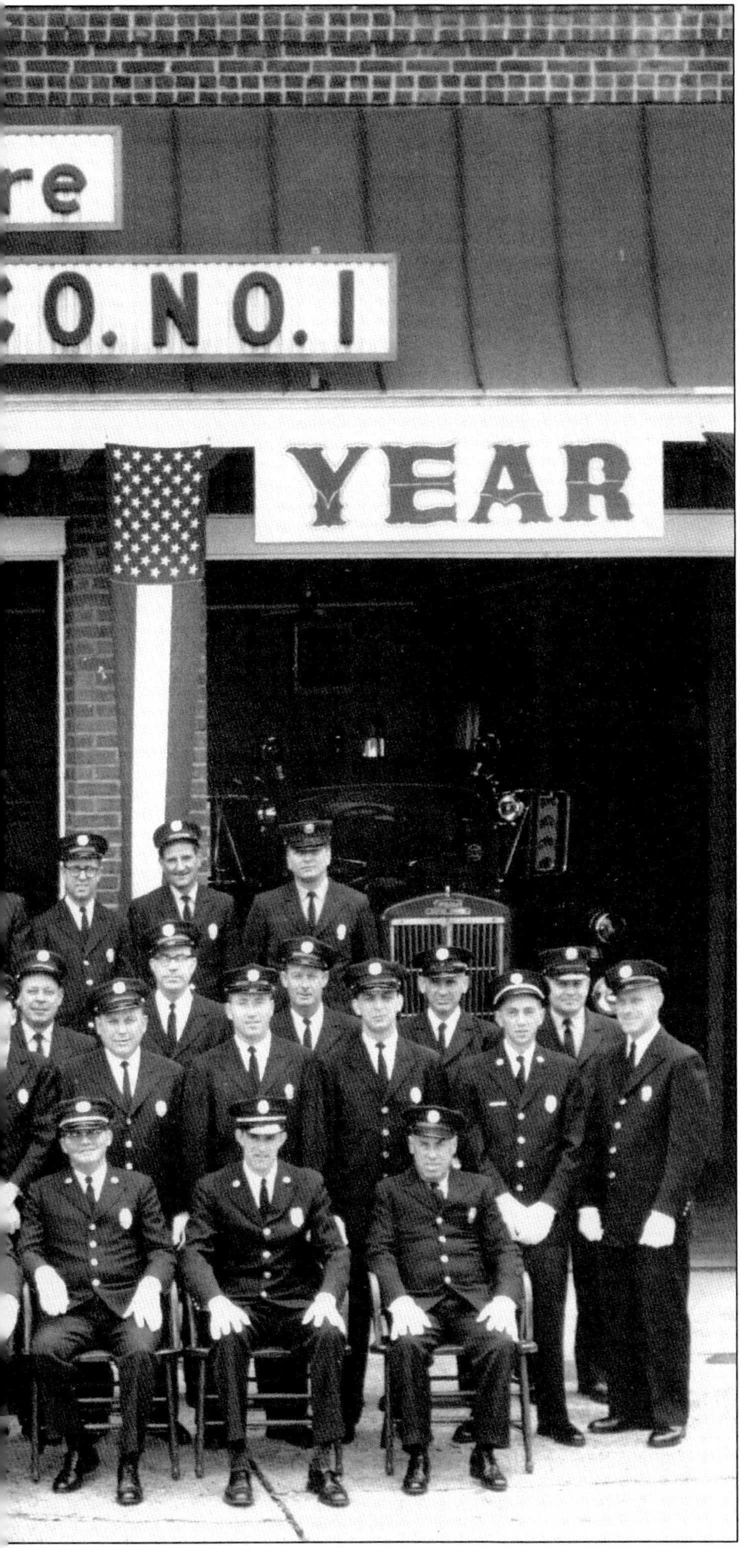

Raritan Engine Company No. 1 celebrated their 50th anniversary in June 1966. Pictured from left to right are (first row) Wendell Slavick, Charles Grand-Jean Jr., Assistant Chief Frances Colletto, chaplain Clifford Voorhees, Chief Eugene Gieber, Frank Rice, Richard Borwegen, and Kenneth Stout; (second row) Anthony Colletto, James Colletto, Curtis Clark, Edward Kashtock, William Colletto, Joseph Stumpf, Leonard Waite, Robert Borwegen, Clarence Latham, Carl Pfieffer, Richard Van Sickle, Dominick DeFrank, John Galambos, and William Borwegen; (third row) Joseph Madarasz, Elwood Waite, George Bogdan, James Ward Sr., Frederick Schulteiss, Peter Bogdan, Roger Vroom, Edward Muckin, Gaza Toth, James Ward Jr., H. Ray Vliet, William Weissenborn, Thomas Downs, Donald Davis, Robert Meluski, James Burke, William Maison, William Stout, and Harold McGorvin; (fourth row) Vincent Anderson, Charles Toth, Bruce Redfield, Herbert Eayres, John Sayben, Peter Borwegen, William Kermes, William Langon, Robert Grand-Jean, William Scheider, George Ellmeyer, and Alexander Danish. (Courtesy of Raritan Engine Company No. 1.)

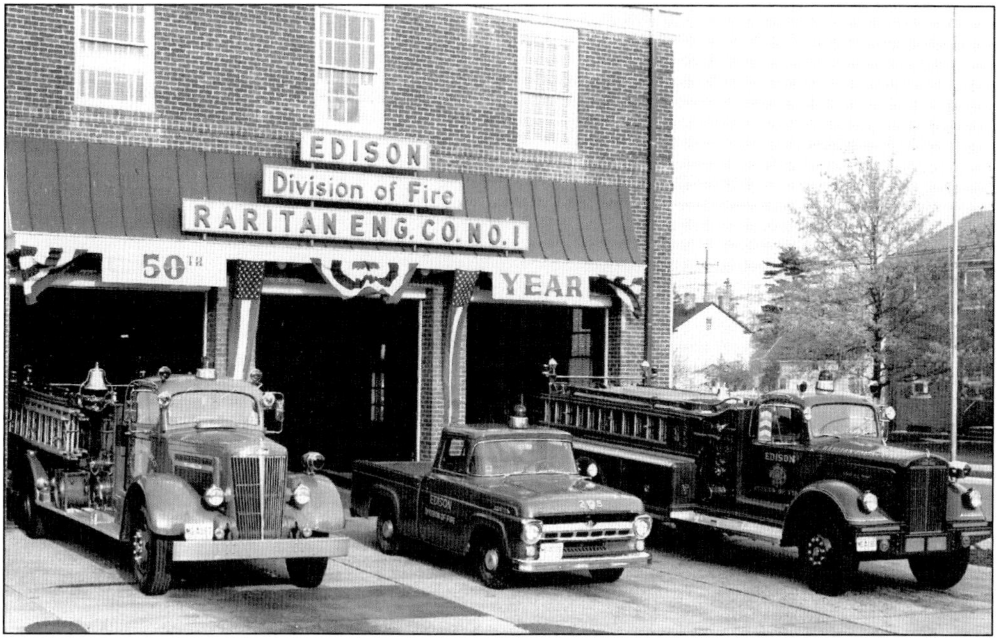

This 1966 photograph shows car 25, a 1957 Ford F100 pick-up truck, centered between engine 1 and engine 2. Engine 1, pictured on the left, was a 1948 White equipped with a 750-gallon-capacity pump. Engine 2, a 1954 White, was also equipped with a 750-gallon pump and was almost identical to engine 1. The fire commissioners purchased both fire engines from the New Jersey Fire Equipment Company in Dunellen. (Courtesy of Raritan Engine Company No. 1.)

RARITAN ENGINE COMPANY NO. 1

EX - CHIEFS

J. H. Becker.............. 1916-17
J. E. Pardun.............. 1918
W. F. Woerner........... 1919
A. W. Latham............ 1920
T. Swales.................. 1921
J. H. Wildgoose........... 1922
P. I. Fisher................ 1923-24
H. K. Drake............... 1925
P. I. Fisher 1926-27
C. E. Stout................ 1928
J. Stalker.................. 1929
J. Horvath................. 1930
R. Hansmann............ 1931
J. Van Stout.............. 1932
E. Rasmussen........... 1933
C. Falger.................. 1934
H. K. Drake............... 1935
W. Slavick................. 1936
T. Eggertson............. 1937
G. H. Graff................ 1938
C. Oliveri.................. 1939
T. Swales.................. 1940
E. Grant................... 1941
E. J. Monaghan......... 1942
E. Rush.................... 1943
E. M. Voorhees.............. 1943
J. Bernat....................... 1944
K. E. Wait...................... 1945
G. S. Van Sickle............ 1946
J. K. Stout...................... 1947
F. Takacs....................... 1948
S. Varga........................ 1949
W. Rush......................... 1950
E. Wunnenberg.............. 1951
R. Ellmyer...................... 1952
R. Bishop....................... 1953
C. Latham...................... 1954
E. Wait.......................... 1955
C. Voorhees................... 1956
W. J. Schneider............. 1957
L. W. Wait...................... 1958
J. Madarasz................... 1959-60
J. Ward......................... 1960-61
E. Muckin....................... 1962
T. Swales III.................. 1963
D. Davis........................ 1964
W. Kermes..................... 1965
E. Geiiber...................... 1966
F. Rice.......................... 1967-68
R. Borwegen.................. 1969
E. Wait...................... 1970
W. Borwegen............ 1971
T. Downs.................. 1972
R. Borwegen............. 1973
P. Borwegen............. 1974
R. Gavin................... 1975
R. Mundy.................. 1976
W. Latham................ 1977
B. Enoch.................. 1978
G. Bogdan................ 1979
J. Tripod.................. 1980-81
R. Campbell.............. 1982-83
B. Latham................. 1984-85
B. Grand Jean........... 1986-87
C. Copperthwaite........ 1988-89
D. Clause.................. 1990-91
D. Borwegen............. 1991-92
E. Hummel................ 1993-94
F. Hohwald............... 1994-95
A. Yourstone............. 1995-97
K. McGorvin.............. 1997-98
B. Szebenyi............... 1998-05
H. Maurath................ 2005-07

Listed above are the chiefs that have led Raritan Engine Company No. 1 throughout the years.

Two

Edison Volunteer Fire Company No. 1

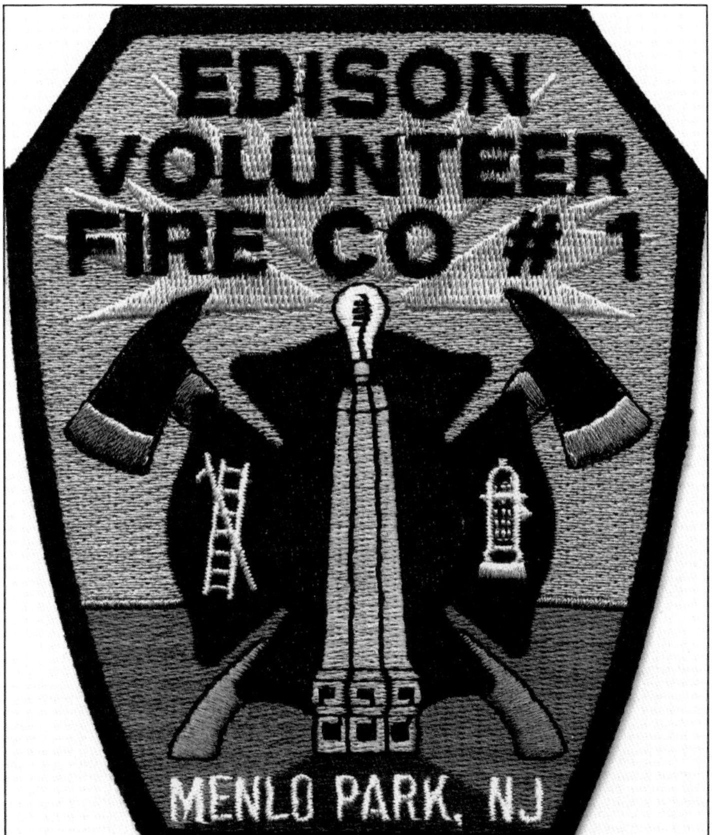

During the mid-1950s, members of the Edison Volunteer Fire Company No. 1 in Menlo Park wanted to wear a patch on their uniforms in order to distinguish themselves from other companies. Fellow member and former chief Mitchell Erceg created the first company patch, in which he drew Thomas Edison's light tower in the center. Today the icon for Edison Township is the light tower, which is displayed on the township seal and on every township department logo.

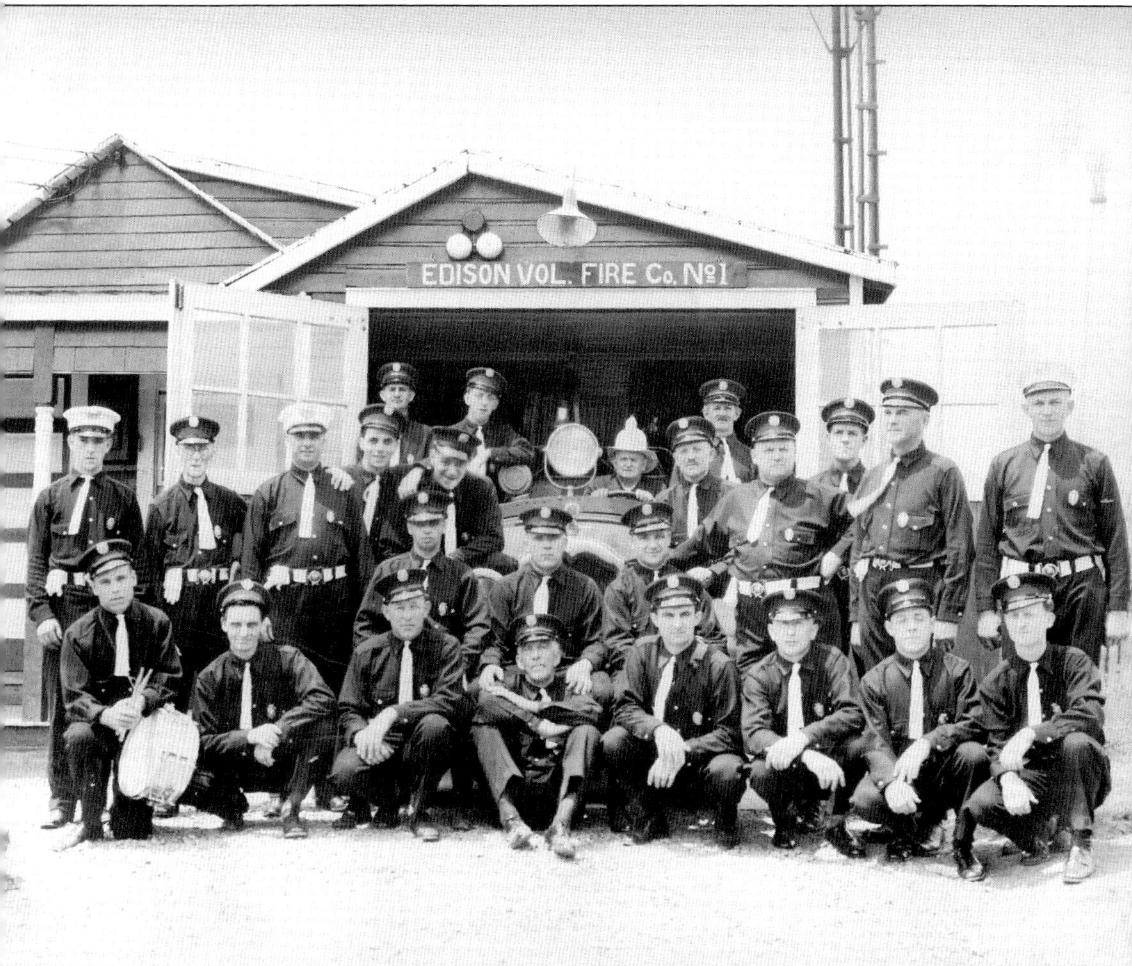

In 1917, a group of passionate Menlo Park residents started providing fire protection by utilizing a horse-drawn wagon equipped with ladders and leather fire buckets. These men housed the wagon in a vacant machine shop at Thomas Alva Edison's old laboratory. Upon being alerted about a fire, fellow firefighter Frederick Peins would bring his horse to the machine shop, hook the horse up to the wagon, and respond to the fire. The company was informally known as the Menlo Park Hook and Ladder Fire Company. On April 4, 1924, Fire District No. 2 and its board of fire commissioners received official recognition from Raritan Township. During this process, the company changed its name to the Edison Volunteer Fire Company No. 1 to honor Menlo Park resident Thomas Alva Edison. Shortly thereafter, local resident Minnie Clarkson sold her Monmouth Avenue bungalow to the company for $500 and the company acquired its first motorized fire engine, Raritan Engine Company No. 1's 1917 Ford Model T, for $26. This 1928 photograph shows the original charter members in front of their newly converted firehouse on Monmouth Avenue. (Edison Volunteer Fire Company No. 1.)

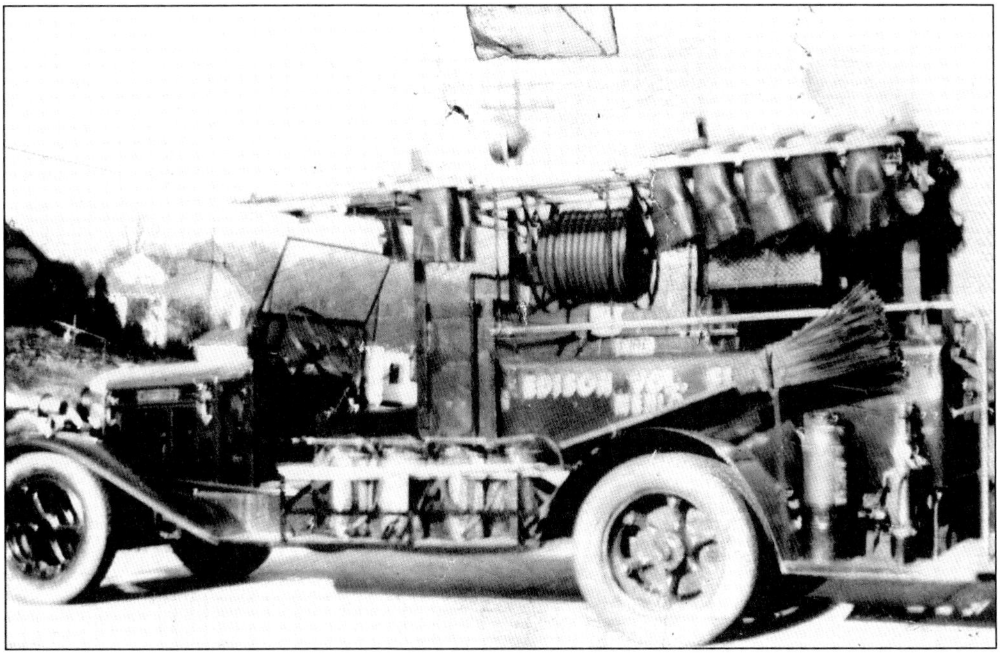

This Diamond-T hose truck was purchased brand new in 1931. It not only carried hose to the scene, but also carried equipment such as brush brooms (for field fires), portable extinguishers, and personal protective gear. In 1939, a pump and booster tank were added and the old Ford Model T was donated to a neighboring town. (Courtesy of Edison Volunteer Fire Company No. 1.)

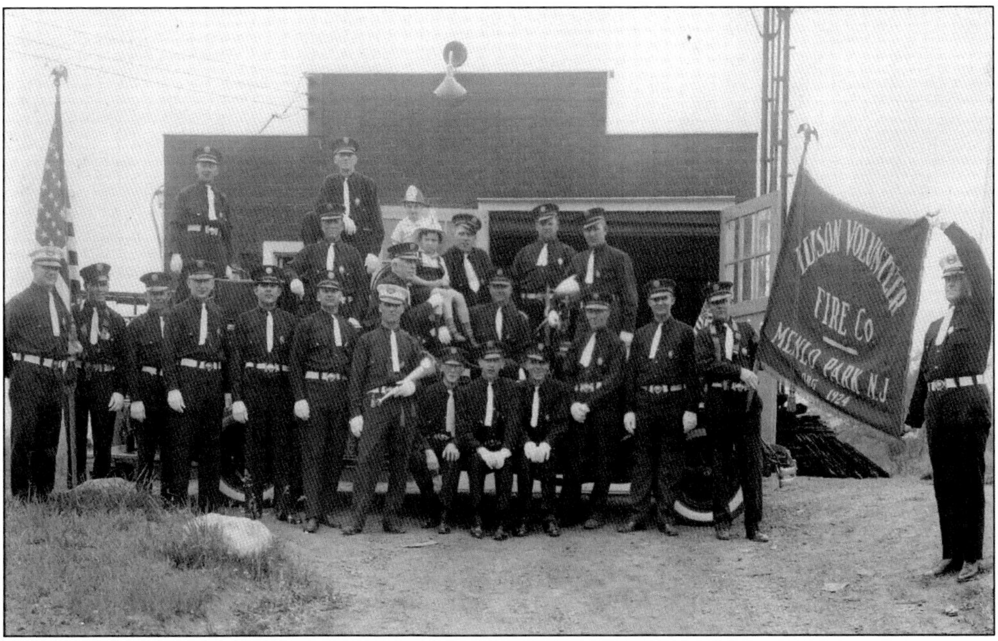

This early-1930s photograph shows members of Edison Volunteer Fire Company No. 1 proudly posing in front of their new Diamond-T hose truck and recently renovated firehouse. The front porch was enclosed and a new brick face was added to the firehouse. (Courtesy of Edison Volunteer Fire Company No. 1.)

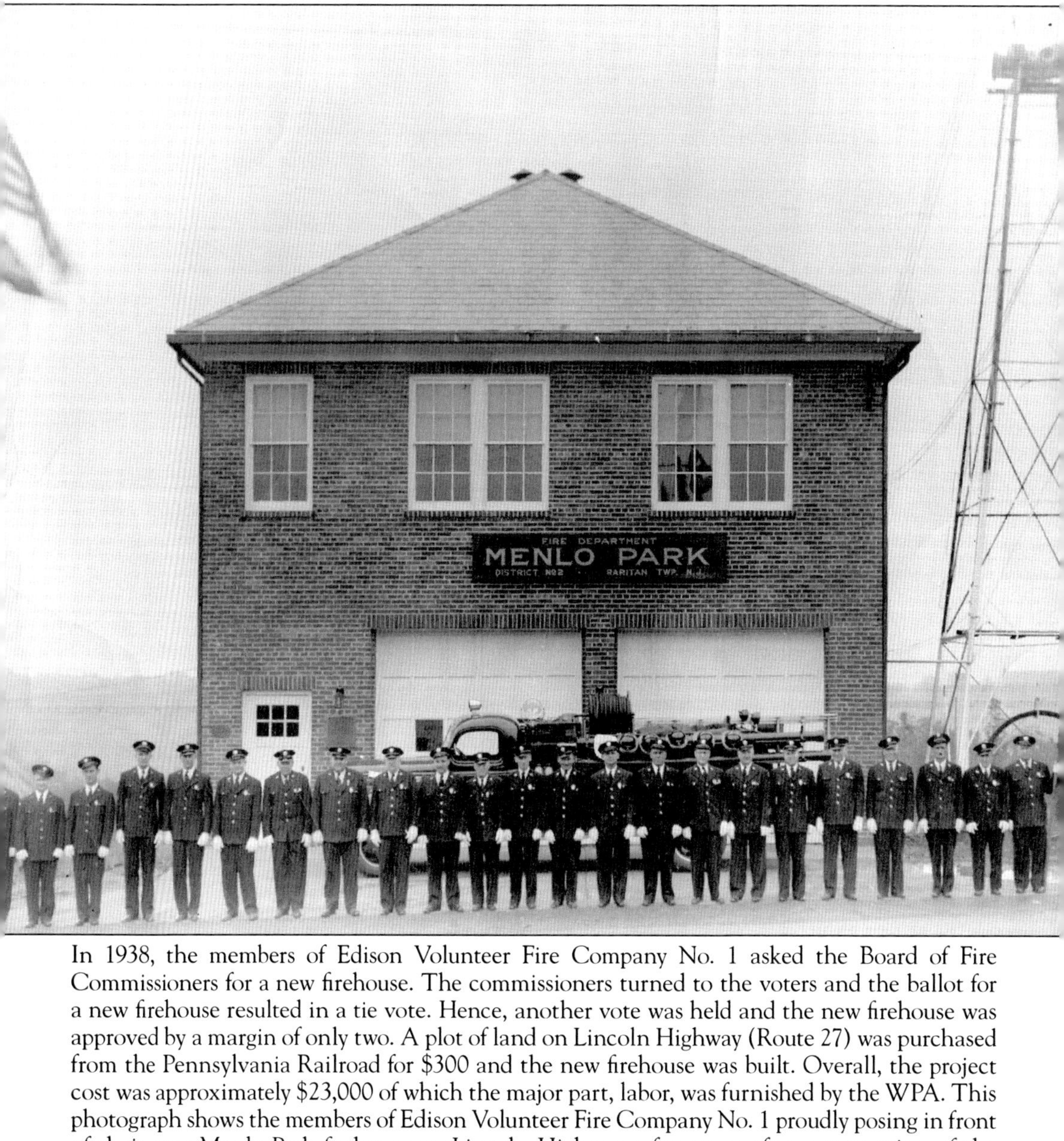

In 1938, the members of Edison Volunteer Fire Company No. 1 asked the Board of Fire Commissioners for a new firehouse. The commissioners turned to the voters and the ballot for a new firehouse resulted in a tie vote. Hence, another vote was held and the new firehouse was approved by a margin of only two. A plot of land on Lincoln Highway (Route 27) was purchased from the Pennsylvania Railroad for $300 and the new firehouse was built. Overall, the project cost was approximately $23,000 of which the major part, labor, was furnished by the WPA. This photograph shows the members of Edison Volunteer Fire Company No. 1 proudly posing in front of their new Menlo Park firehouse on Lincoln Highway a few years after construction of the building. The apparatus behind the members is a 1942 GMC surplus fire engine that was capable of pumping 500 gallons of water per minute. Also, note the old horn and siren tower with the iron ring-shaped gong near the base. (Courtesy of Edison Volunteer Fire Company No. 1.)

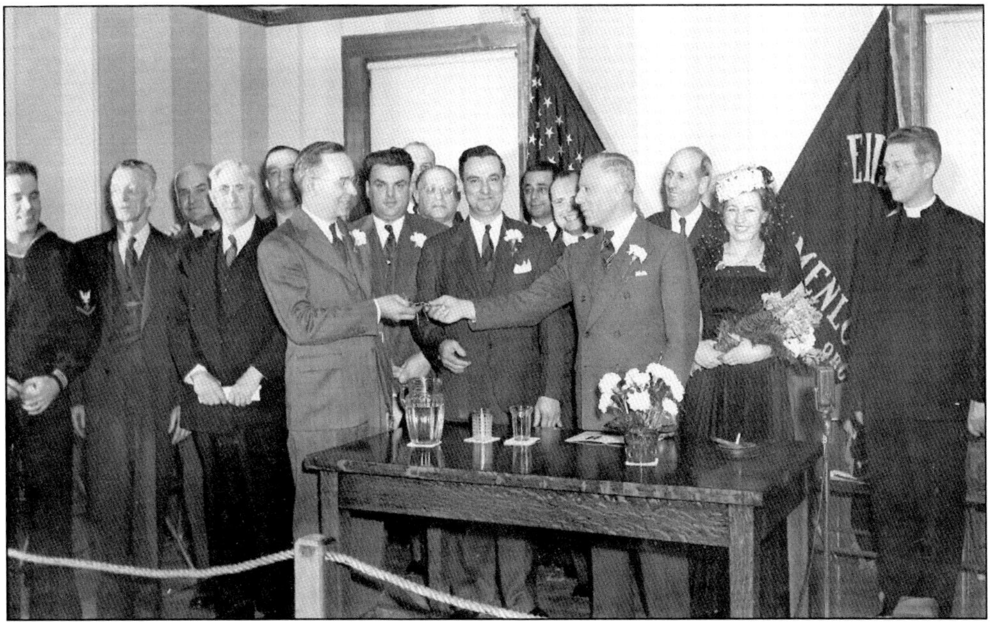

This photograph shows Jacob Weitzen, director of the WPA, handing the building keys over to Fire Commissioner William Sorg during the firehouse dedication ceremony on April 12, 1941. Other honorary guests include, from left to right, former commissioners John Wilkens and Edward Slade, township commissioner John Pardun, architect Charles Carman, sheriff Julius Engel, commissioner Kenneth Shepard, commissioner Alfred Schnebbe, township commissioner James Forgione, and superintendent of Menlo Park Soldiers' Home Maj. George Giger. Lucy Grieb sang the national anthem and Rev. Harold Dunne gave the invocation. (Courtesy of Edison Volunteer Fire Company No. 1.)

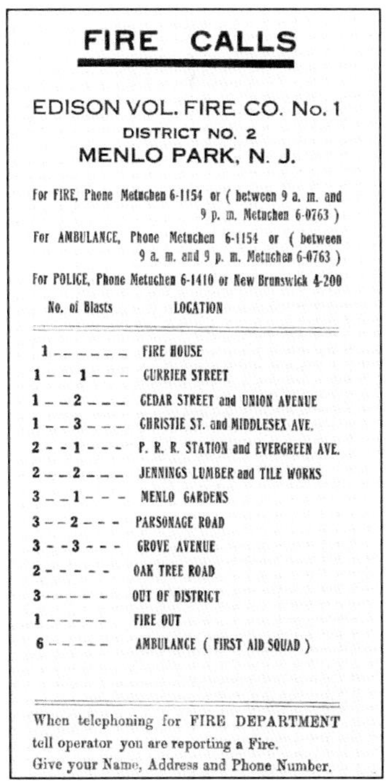

FIRE CALLS

EDISON VOL. FIRE CO. No. 1
DISTRICT NO. 2
MENLO PARK, N. J.

For FIRE, Phone Metuchen 6-1154 or (between 9 a. m. and 9 p. m. Metuchen 6-0763)

For AMBULANCE, Phone Metuchen 6-1154 or (between 9 a. m. and 9 p. m. Metuchen 6-0763)

For POLICE, Phone Metuchen 6-1410 or New Brunswick 4-200

No. of Blasts	LOCATION
1	FIRE HOUSE
1 - - 1	CURRIER STREET
1 - - 2	CEDAR STREET and UNION AVENUE
1 - - 3	CHRISTIE ST. and MIDDLESEX AVE.
2 - - 1	P. R. R. STATION and EVERGREEN AVE.
2 - - 2	JENNINGS LUMBER and TILE WORKS
3 - - 1	MENLO GARDENS
3 - - 2	PARSONAGE ROAD
3 - - 3	GROVE AVENUE
2	OAK TREE ROAD
3	OUT OF DISTRICT
1	FIRE OUT
6	AMBULANCE (FIRST AID SQUAD)

When telephoning for FIRE DEPARTMENT tell operator you are reporting a Fire.
Give your Name, Address and Phone Number.

Before the era of cellular telephones and electronic pagers, firefighters would respond to certain locations based on the number of blasts from the station's horn. This dispatch card, which dates back to the 1940s, shows locations corresponding to the number of blasts. Some landmarks, such as the Pennsylvania Railroad station near Evergreen Avenue, Jennings Lumber Yard, Tile Works, and Menlo Gardens, are no longer in existence and Currier Street has been renamed Wood Avenue. (Courtesy of Morris Kaplan.)

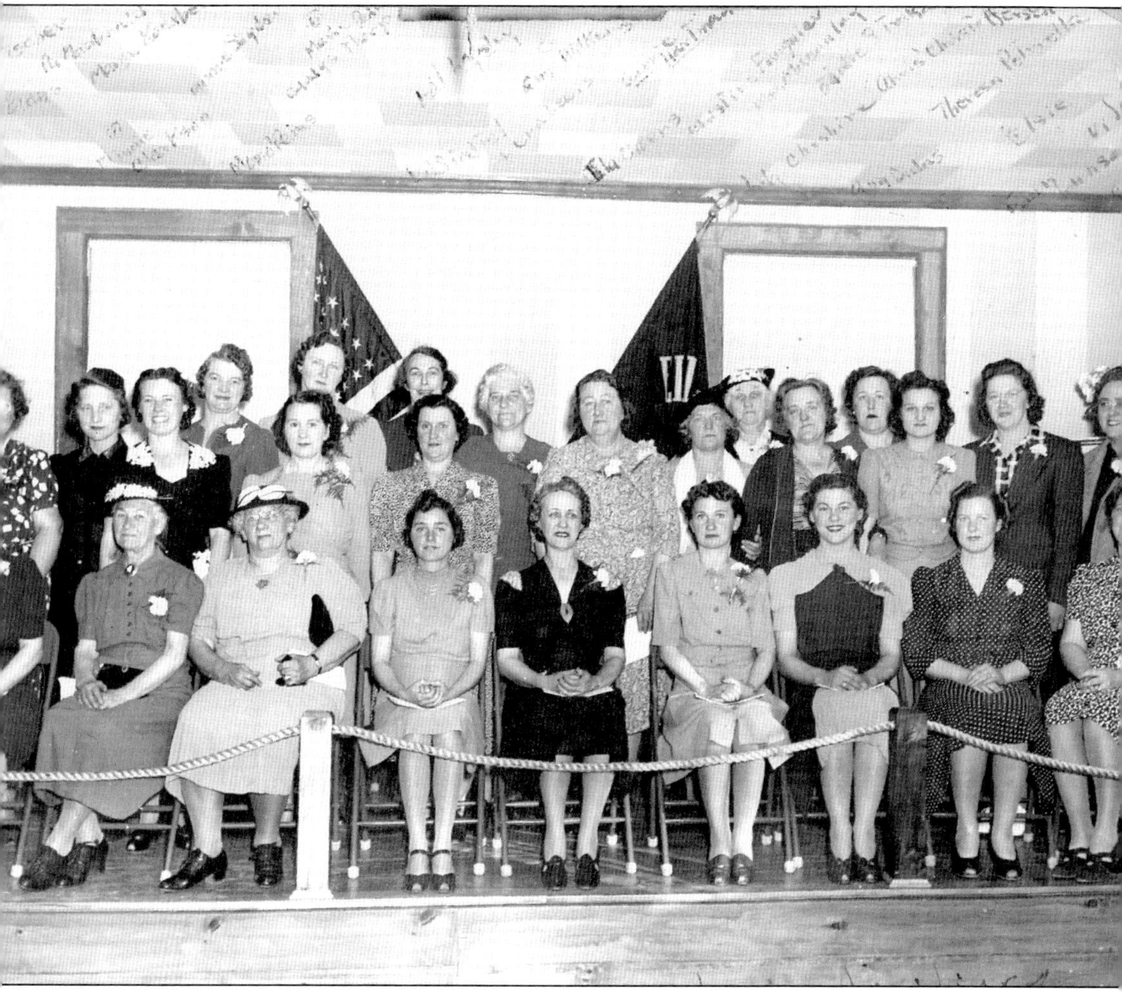

The idea for a ladies auxiliary was conceived during a monthly membership meeting of the Edison Volunteer Fire Company No. 1 at the old firehouse on Monmouth Avenue. On March 28, 1941, a group of women received their charter and became officially known as the Ladies Auxiliary of the Edison Volunteer Fire Company No. 1 of Menlo Park. The auxiliary not only supported the firefighters at emergency scenes, but also organized various fund-raisers and social events. Pictured from left to right are (first row) Sally Evans, Minnie Clarkson, Mamo Peins, treasurer Lillian Stadtel, vice president Sla Owens, president Lilian Cheshire, secretary Anne Dudas, Francis Johnson, and Claire Schnebbe; (second row) Dorothy Fisher, Petra MacDonald, Margaret Koerber, Minnie Snyder, Gladys Thorpe in front of Marie Peins, Lena Lewis in front of Lillian Lapsley, Arma Wilkens, Carrie Hartman, Elizabeth Sorg, Ella Macavley, Bertha Straka, Chris Christoffersen, Theresa Petriella, Elsie MacFarlane, and Viola Jennings. (Courtesy of Edison Volunteer Fire Company No. 1.)

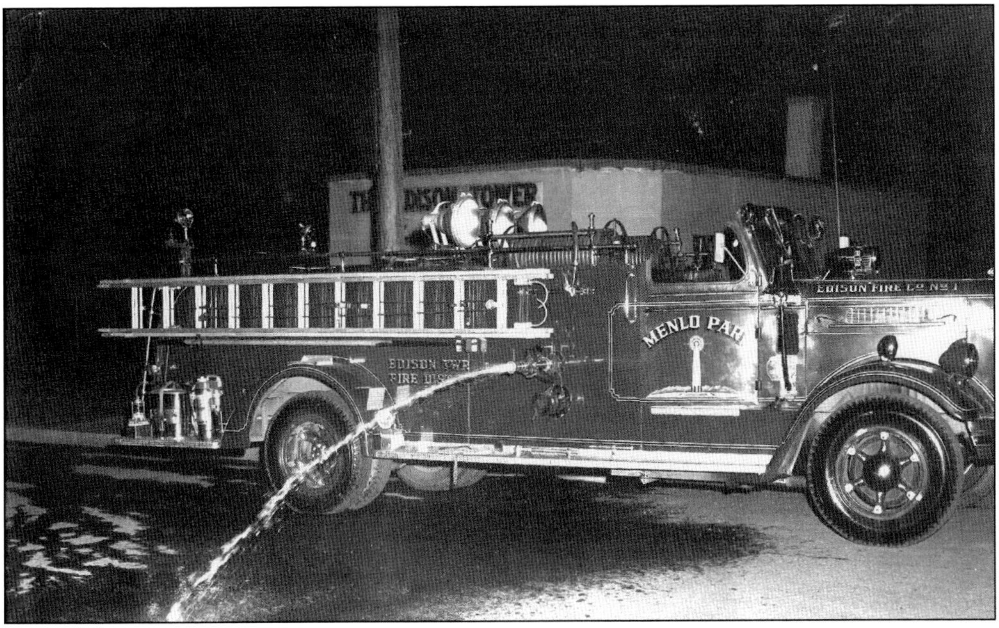

In 1953, the voters of Fire District No. 2 approved a $15,000 bond issue for the purchase of a new fire engine. A 1953 open-cab Oren was received later that year with the emblem of Edison's light tower painted on each door. This photograph of the Oren flowing water was taken in front of the Edison Tower Museum on Christie Street. (Courtesy of Edison Volunteer Fire Company No. 1.)

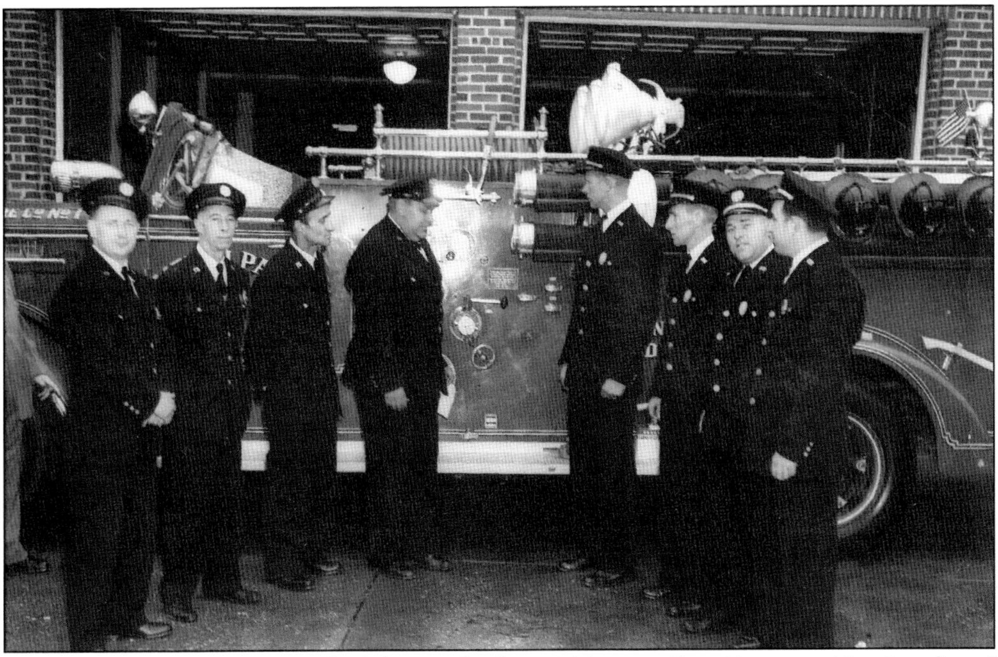

This 1953 photograph shows firefighters of Edison Volunteer Fire Company No. 1 gathered together in front of their new Oren fire engine. Pictured from left to right are Martin Dige, unidentified, unidentified, Mac McGraff, Chief Lee Sofield, Robert Gray, Earl Evans, and William Evans. (Courtesy of Edison Volunteer Fire Company No. 1.)

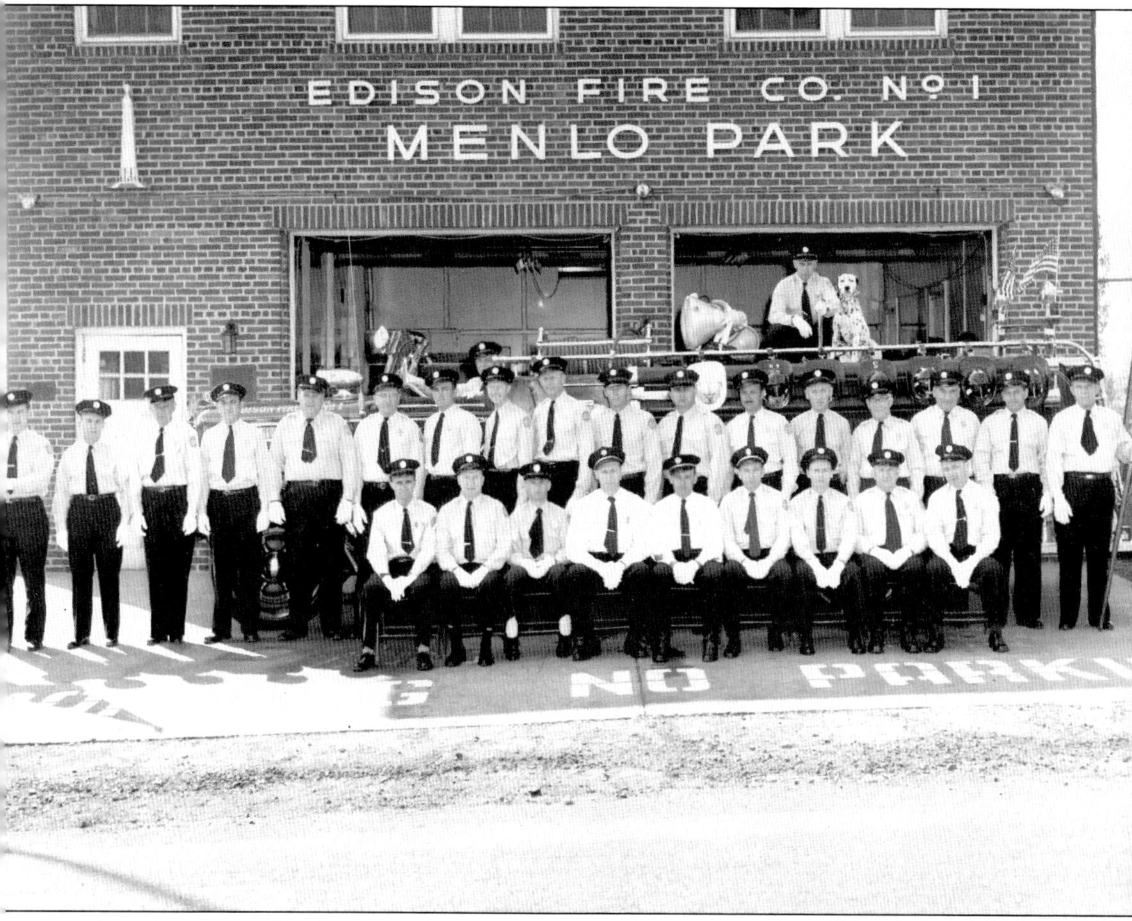

Edison Volunteer Fire Company No. 1 members gather together for a formal photograph on Memorial Day 1958. This photograph shows the new lettering on the firehouse, which replaced the old Raritan Township sign. This was the year that Edison Township had undergone a major change in government, which resulted in the abolishment of fire districts and fire commissioners. Notice that each member wore the new company patch on his left shoulder. This fire company was not only the first volunteer fire company in Edison to have a company patch, but also the first organization to use Thomas Edison's name and light tower in his honor. The light tower is now an icon for Edison Township. Pictured from left to right are (first row) unidentified, unidentified, Malcolm Thorn Jr., Arnold Hergenhan, Chief Mitchell Erceg, Joseph Hagerty, Harold Willis, Peter Thorpe, and Martin Dige; (second row) Herman Good, unidentified, Steward Straka, Walter Konesky, Mac McGraff, unidentified, unidentified, Charles Kohlbusch, Lee Sofield, Carl Otterson, unidentified, William Evans, Robert Gray, John MacDonald, unidentified, Louis Gurdon, and Rudolph Peins. John Wilkens is sitting behind the wheel and Donald Yackel is posing up on top of the hose bed with his dalmatian, Patches. (Courtesy of Edison Volunteer Fire Company No. 1.)

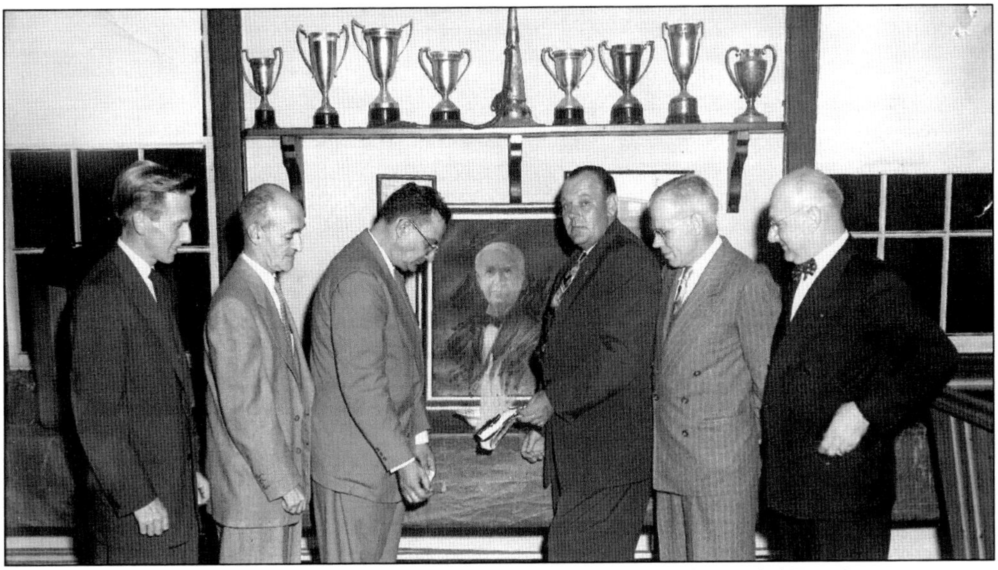

This photograph captures a momentous occasion in which (from left to right) Fire Commissioners Robert Gray, Virgil Owens, Albert Schnebbe, Steward Straka, John Wilkens, and Henry Koerber burn the firehouse mortgage in the early 1950s. Notice the portrait of Thomas Edison in the background. This oil painting was given to the company from the Edison Pioneers and Edison Laboratories of West Orange during the firehouse dedication on April 12, 1941. The painting is now on display at the Edison Municipal Complex. (Courtesy of Edison Volunteer Fire Company No. 1.)

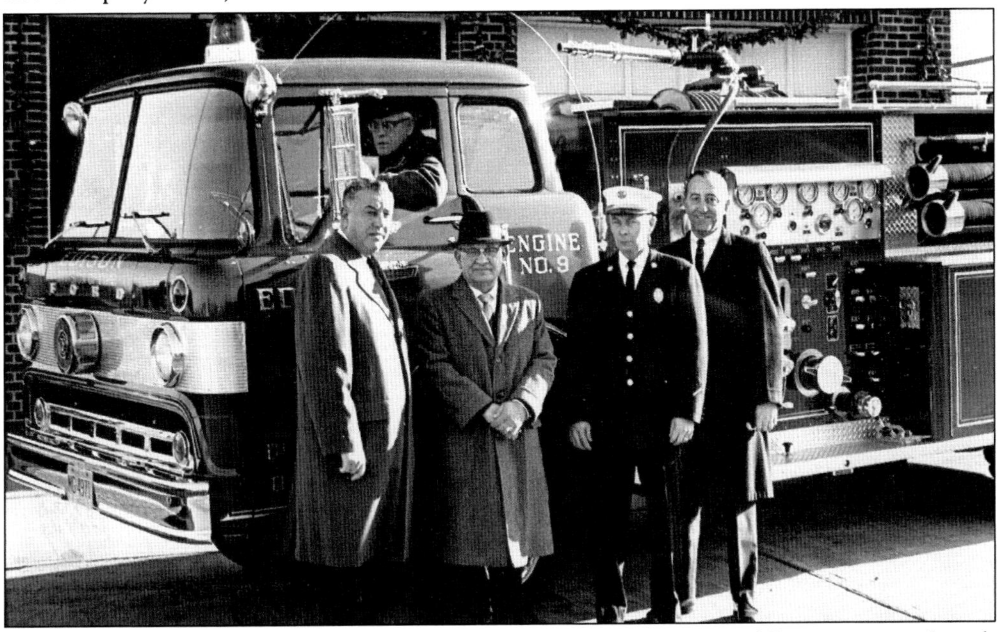

Seen from left to right, Mayor Anthony Yelencsics, Edison Division of Fire supervisor Joseph Simon, volunteer chief Donald Yackel, and business administrator John Delasandro pose in front of a new 1966 Ford pumper. Edison volunteer firefighter John Wilkins (behind the wheel) personally inspected the engine and drove it back to Menlo Park from Missouri. (Courtesy of Edison Volunteer Fire Company No. 1.)

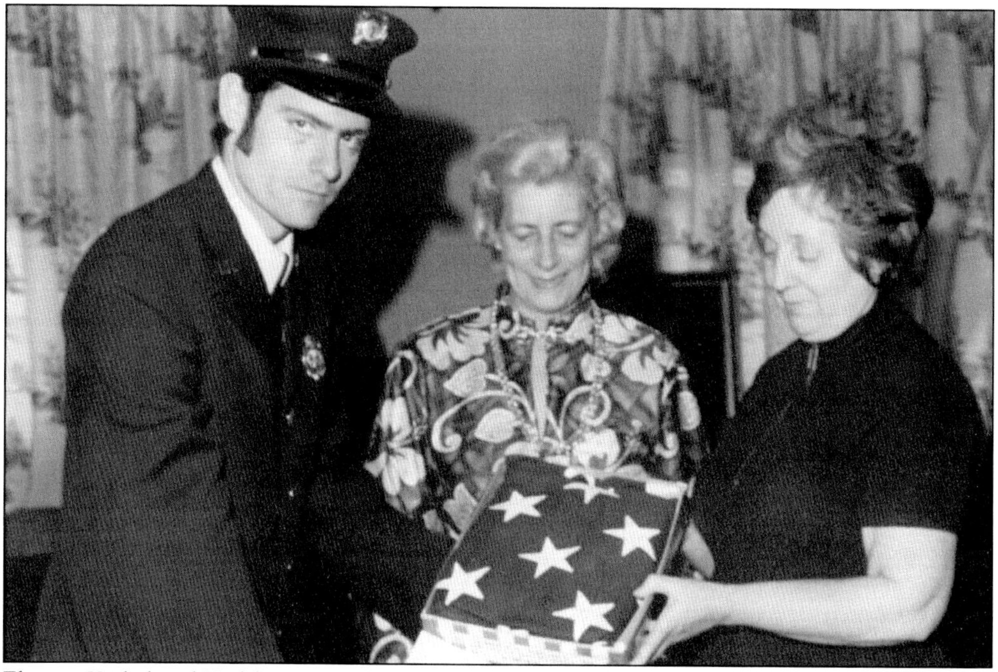

Elenore Yackel and Kathleen Campbell, on behalf of the ladies auxiliary, presented an American flag to the firefighters of Edison Volunteer Fire Company No. 1 on February 13, 1972. Accepting the gift is firefighter Stan Cohen. The flag, which had been obtained from United States representative Edward J. Patten, flew above the nation's capital. (Courtesy of Edison Volunteer Fire Company No. 1.)

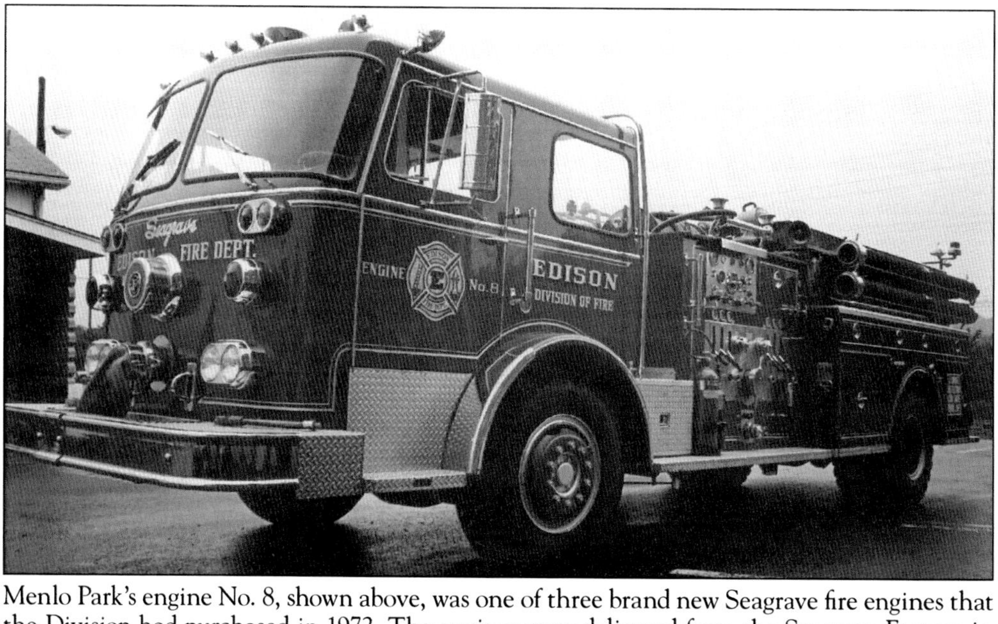

Menlo Park's engine No. 8, shown above, was one of three brand new Seagrave fire engines that the Division had purchased in 1972. The engines were delivered from the Seagrave Factory in Wisconsin six months later than projected. The engines were purchased for $52,000 each and were the first diesel-powered engines ordered by the Edison Division of Fire. (Courtesy of Edison Volunteer Fire Company No. 1.)

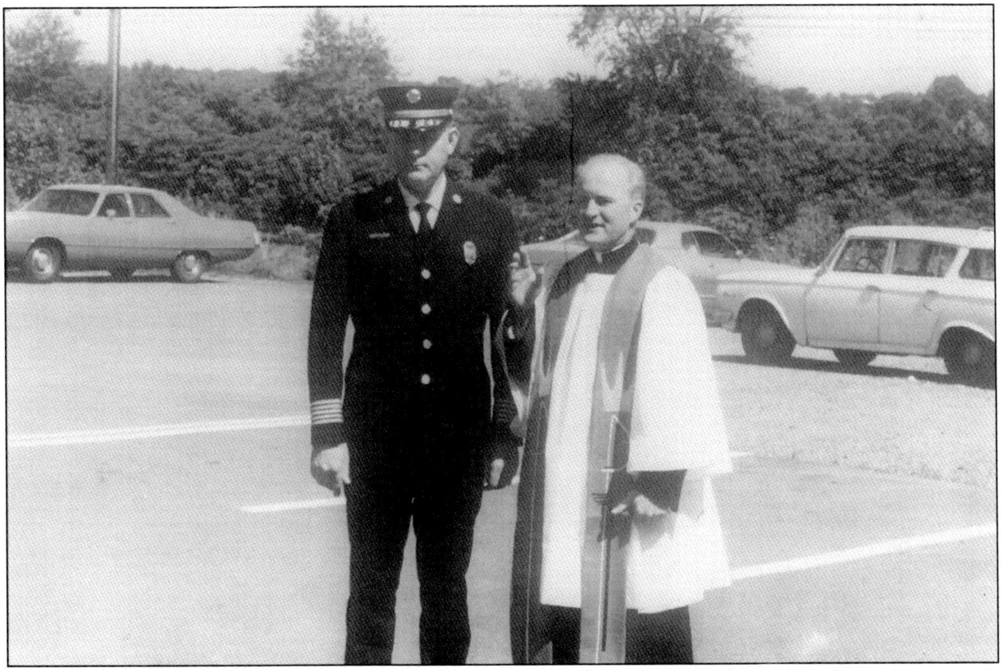

Edison Division fire chief H. Ray Vliet and Rev. Armando Perini stand together for a photograph during the dedication ceremony of the new 1972 Seagrave fire engine. Perini has been a pastor of St. Helena's Roman Catholic Church on Grove Avenue since 1972 and was honored with the title of reverend monsignor in 1996. (Courtesy of H. Ray Vliet.)

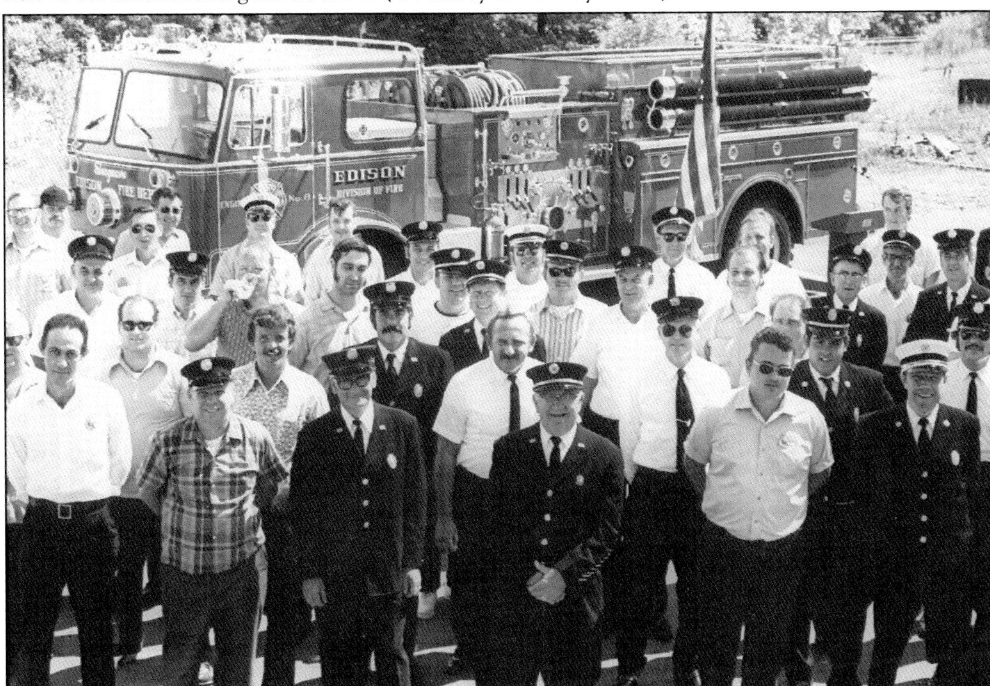

The 1972 dedication of the Edison Volunteer Fire Company No. 1's new fire engine attracted firefighters from all over. (Courtesy of Edison Volunteer Fire Company No. 1.)

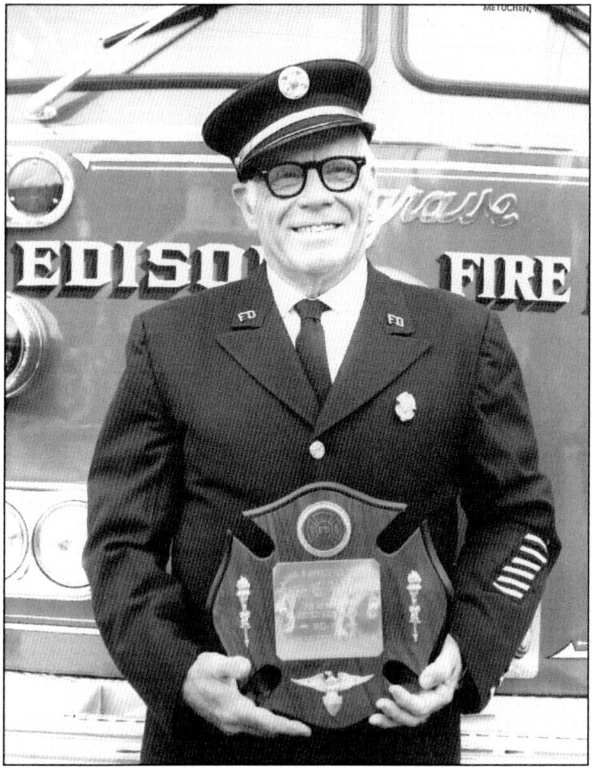

In this photograph, volunteer firefighter John Wilkens poses with his new plaque citing 40 years of service. Edison Volunteer Fire Company No. 1 honored Wilkens with this plaque on July 13, 1973, for his dedication to the company. During the ceremony, Wilkens referred to the Jennings Lumberyard fire as one of the worst fires that he had seen, adding that it took almost five hours to extinguish. He also reminisced about his childhood and his early desire to serve the fire company. His father, a former assistant chief and original charter member of the company, inspired Wilkens at an early age. (Courtesy of Edison Volunteer Fire Company No. 1.)

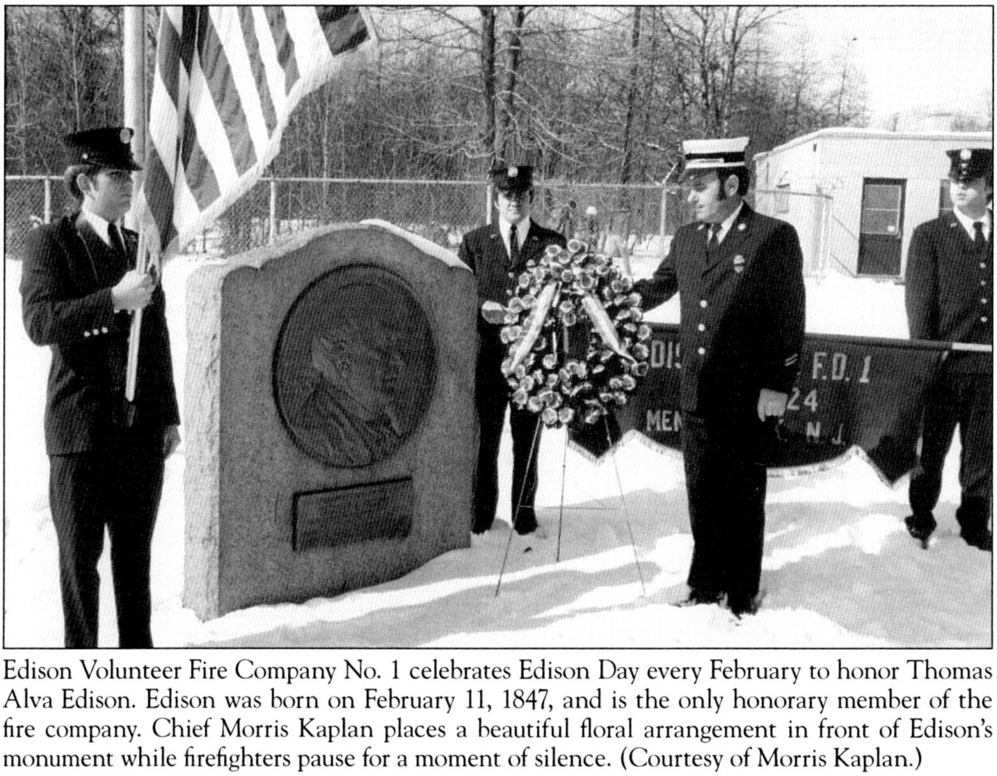

Edison Volunteer Fire Company No. 1 celebrates Edison Day every February to honor Thomas Alva Edison. Edison was born on February 11, 1847, and is the only honorary member of the fire company. Chief Morris Kaplan places a beautiful floral arrangement in front of Edison's monument while firefighters pause for a moment of silence. (Courtesy of Morris Kaplan.)

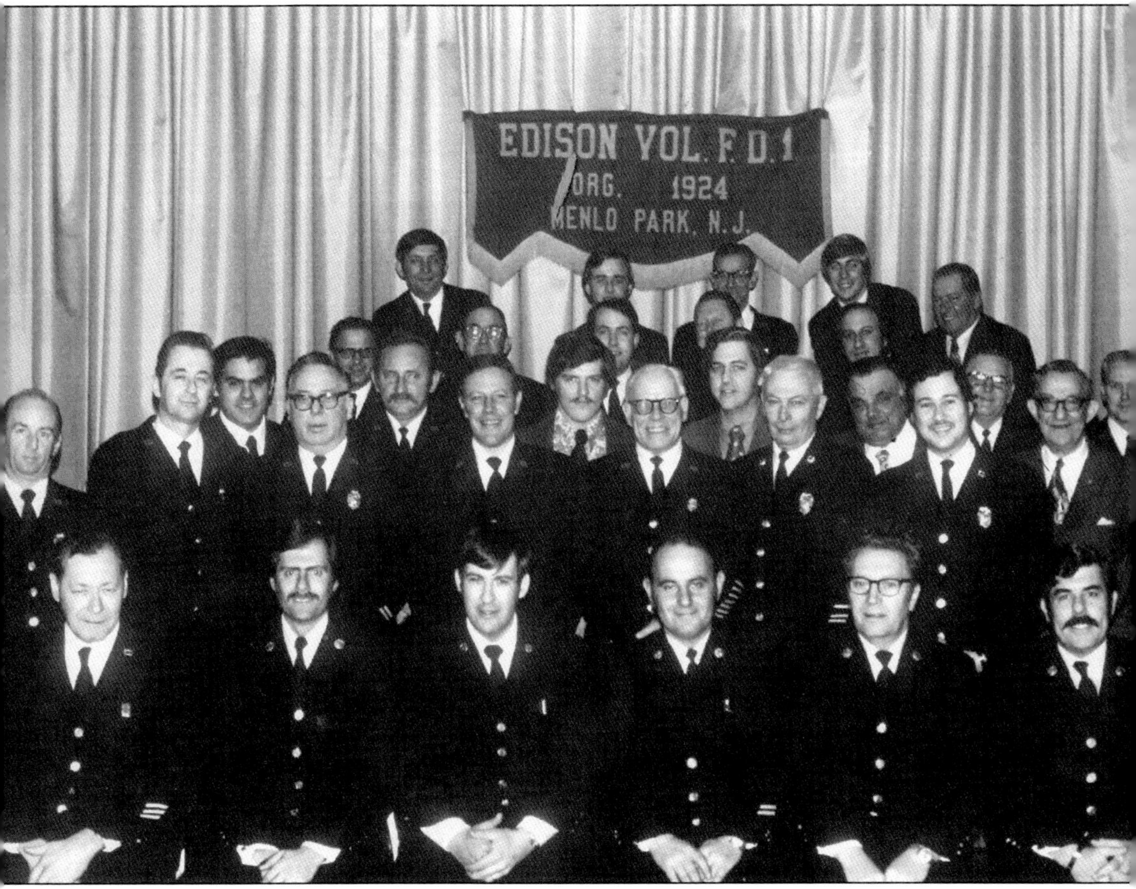

Members of Edison Volunteer Fire Company No. 1 gather together for a group picture during the 1974 Fire Officer Installation Dinner. Every year the company elects new officers and a special ceremony and dinner is held at a local catering hall. Pictured from left to right are (first row) Chief Engineer Arnold Hergenhan, Assistant Chief Stanley Cohen, Allen Francis, Chief Morris Kaplan, Joseph Martin, Capt. Victor D'Addario; (second row) Malcolm Thorne Jr., Robert Wagner, James Shephard, Joseph Scott, James Campbell, John Guilfoyle, unidentified, John Wilkens, Raymond Bostard, Donald Yackel, Kenneth Shephard, Neil Kenny, George Taylor Sr., unidentified, and unidentified; (third row) Louis Gurdon, Edward Konesky, G. Robert Campbell, Charles Kohlbusch, and Michael Mermelstein; (fourth row) Norman Jensen Sr., Barry Thorne, George A. Campbell, Robert Yackel, and unidentified. (Courtesy of Edison Volunteer Fire Company No. 1.)

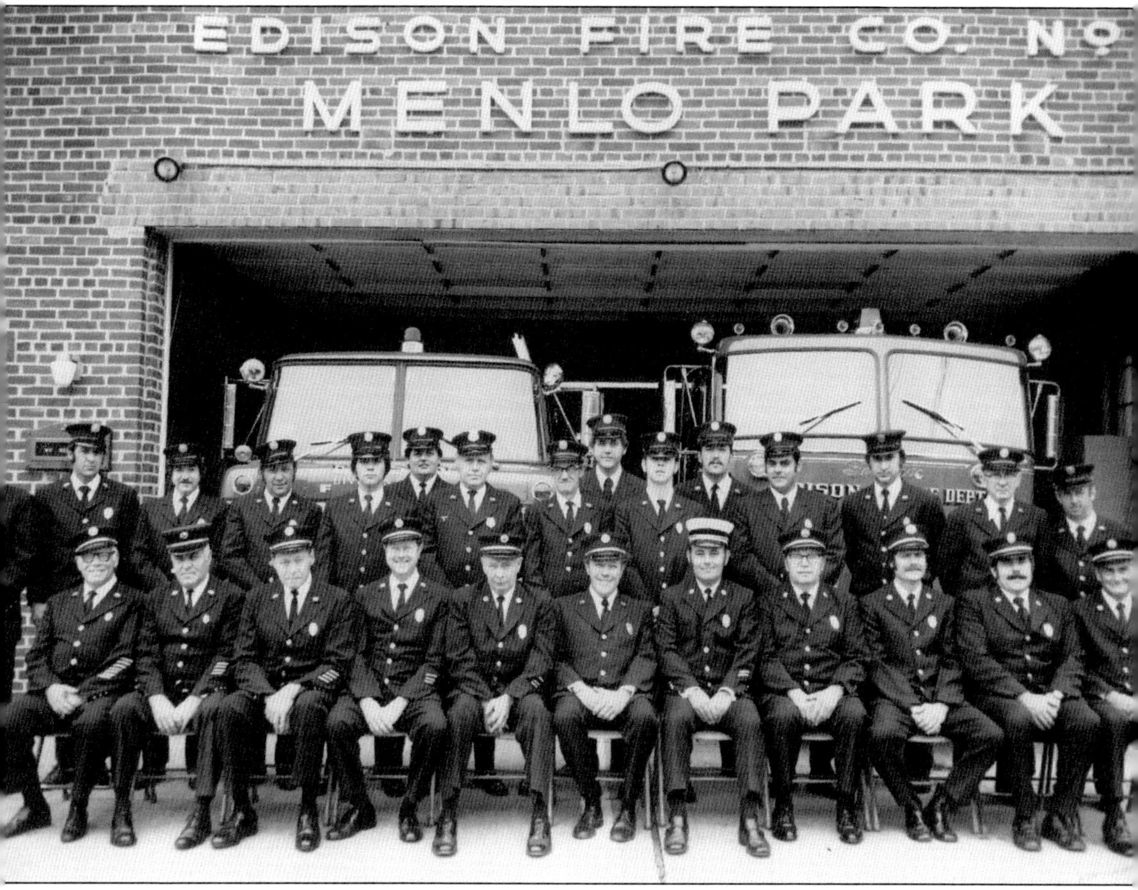

Decoration Day, now known as Memorial Day, was first observed on May 30, 1868, when flowers were placed on the graves of Civil War soldiers at Arlington National Cemetery. This tradition of remembrance continued annually on the 30th day of May until 1971, when Congress passed the National Holiday Act, which changed the date to the last Monday of May to ensure a three-day weekend. Recently, many fire companies have adopted the last weekend in May to honor and remember their departed members. This photograph shows the members of Edison Volunteer Fire Company No. 1 posing in front of the firehouse on Memorial Day 1974. Pictured from left to right are (first row) John Wilkens, Wesley Wilkens, Arnold Hergenhan, Charles Kohlbusch, Donald Yackel, John Guilfoyle, Chief Morris Kaplan, Joseph Martin, Stanley Cohen, Victor D'Addario, and Walter Katula; (second row) Neil Kenny, Allen Francis, Allan Blander, Norman Jensen Sr., Norman Jensen Jr., James Mulvey, John Spitler, George Campbell, Raymond Bostard, Barry Thorne, G. Robert Campbell, James Shepard, Michael Mermelstein, Hugh Grapes, and Malcolm Thorne Jr. (Courtesy of Morris Kaplan.)

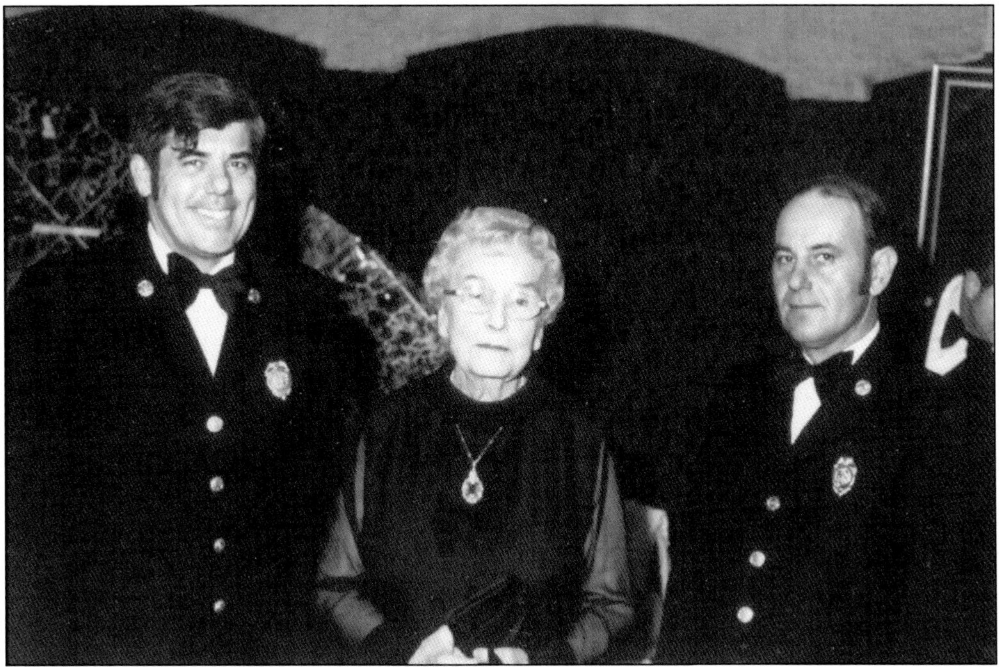

Firefighter Charles Kohlbusch (left) and Chief Morris Kaplan welcome Madeline Edison, daughter of Thomas Alva Edison, to the Edison Volunteer Fire Company No. 1's golden anniversary celebration at the Pines Manor on April 20, 1974. (Courtesy of Morris Kaplan.)

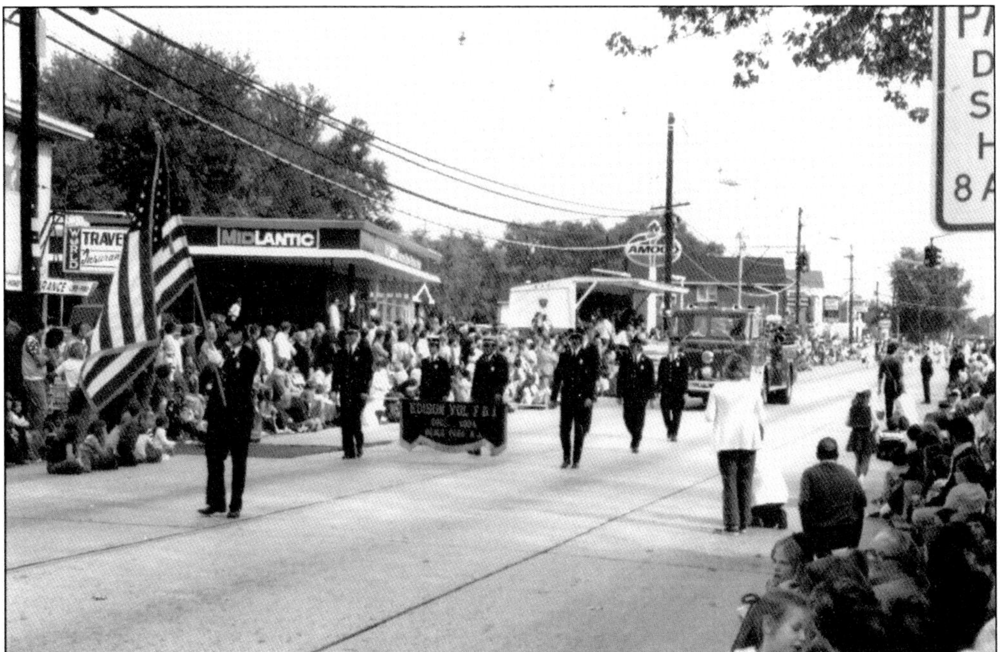

Edison residents were excited to see Menlo Park's volunteer fire company, led by Chief Morris Kaplan, marching down Amboy Avenue during the 1974 Memorial Day parade. (Courtesy of Morris Kaplan.)

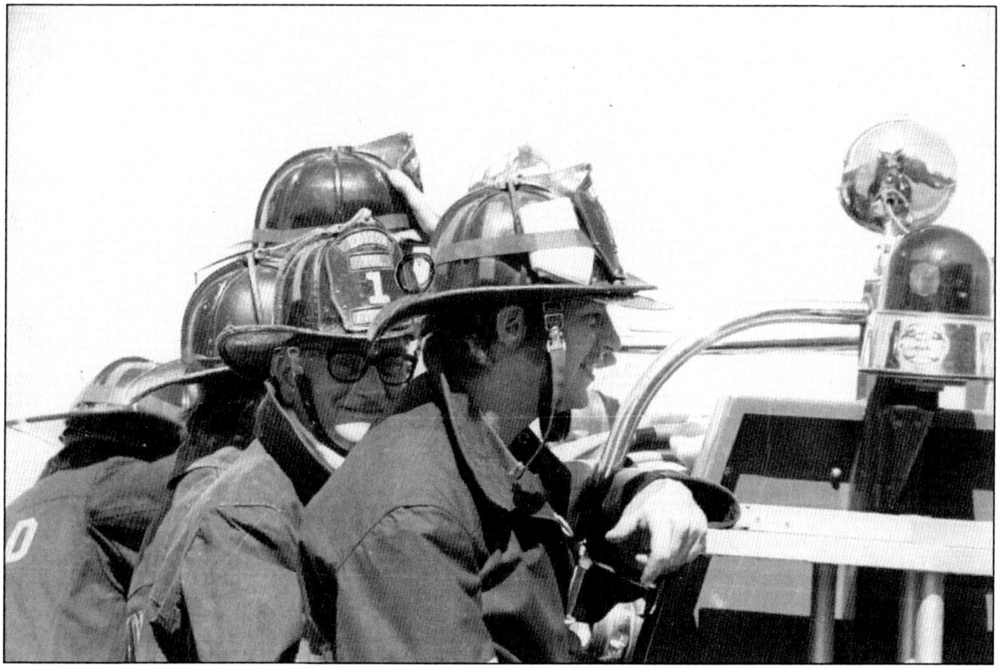

When Menlo Park's alarm sounded, firefighters would quickly respond to the firehouse in order to "catch" the engine before it pulled out. It was common for firefighters to hitch a ride on the engine's tailboard en route to the fire because there were no seats available. This dangerous practice stopped in the mid-1980s. (Courtesy of Thomas Walsh.)

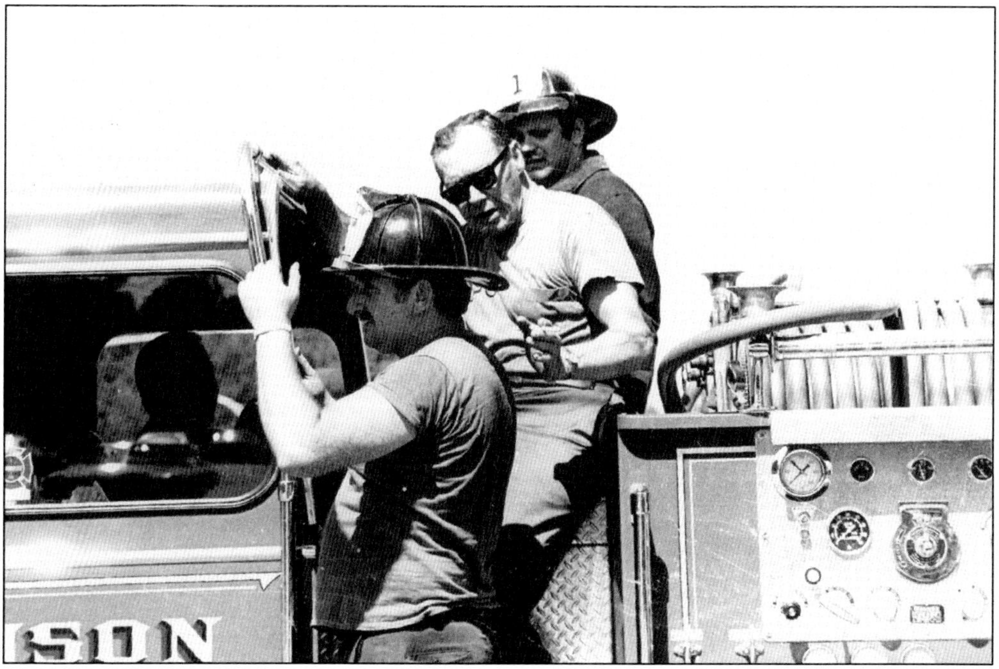

This 1977 photograph captures firefighters James Campbell, Arnold Hergenhan, and Walter Boychick discussing their previous fire call while returning to the firehouse. (Courtesy of Thomas Walsh.)

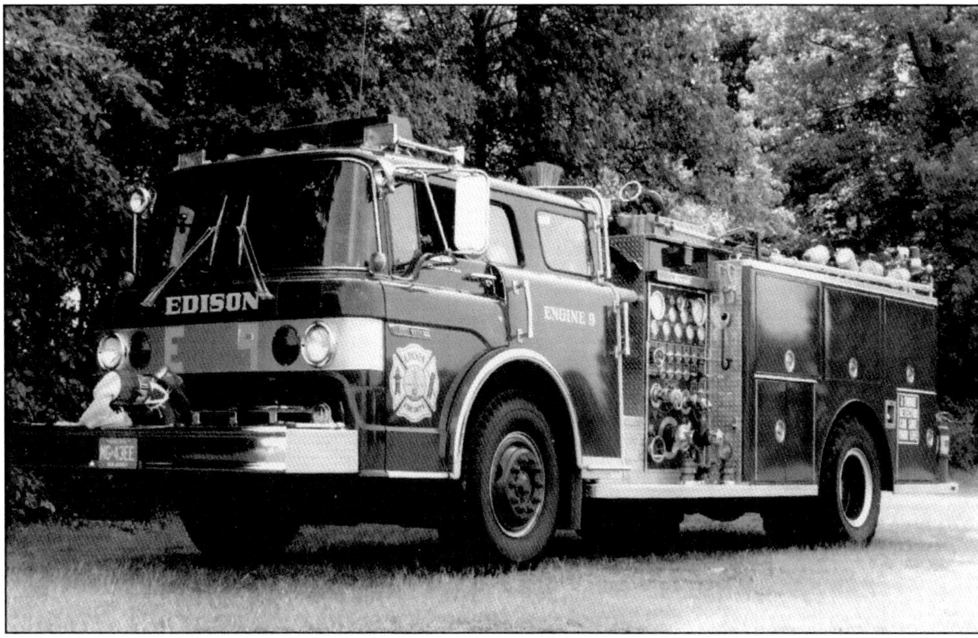

The Edison Division of Fire purchased two new Ford fire engines in 1985. One engine became engine No. 4, which was housed at Station No. 2 on Route 27 in the southern part of Edison and the other, which is shown above, replaced Menlo Park's 1966 Ford and became engine No. 9. (Courtesy of Thomas Walsh.)

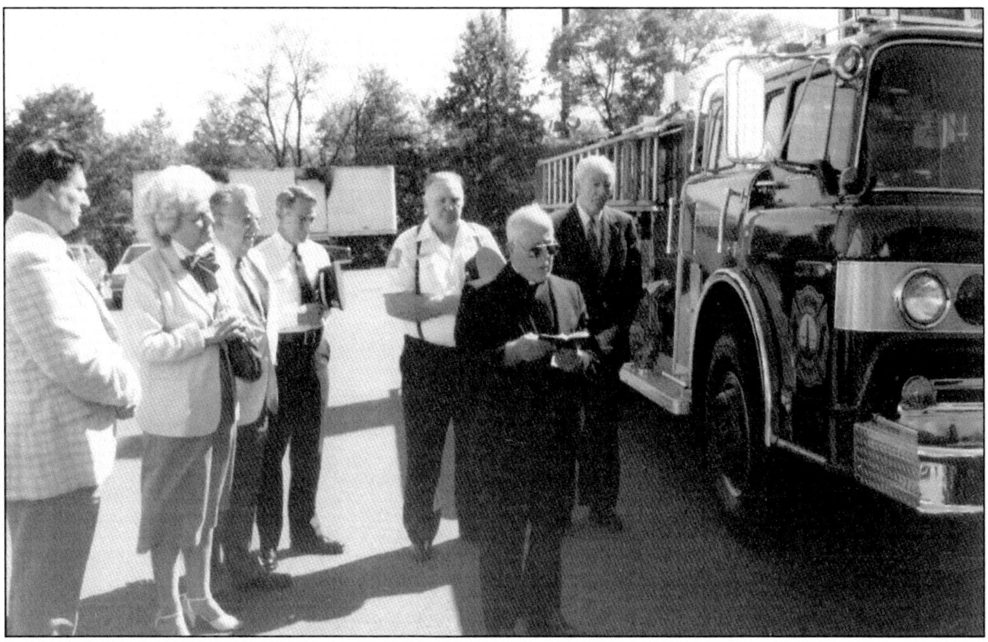

Rev. Joseph Rossetti of St. Helena's Roman Catholic Church is offering a special blessing during the dedication ceremony of the new 1985 Ford fire engine. Standing behind Rossetti are, from left to right, Frank Nussant, council member Dorothy Drwal, council member Sidney Frankel, Division of Fire chief Richard Latham, firefighter George Taylor Sr., and Mayor Anthony Yelencsics. (Courtesy of Edison Volunteer Fire Company No. 1.)

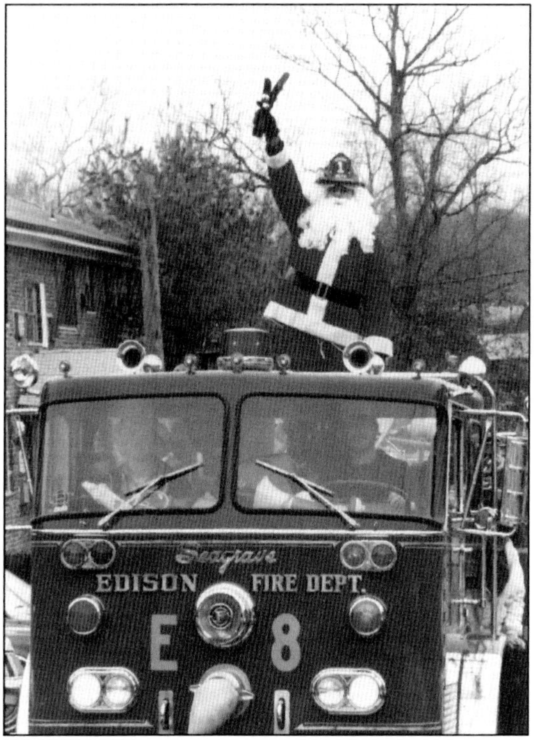

It has been tradition for the members of Edison Volunteer Fire Company No. 1 to escort Santa Claus on the fire engine throughout the precinct. The siren on the slow-moving fire engine calls children from their homes to receive a small treat from Santa's elves. Santa also visits the children's ward of John F. Kennedy Medical Center on James Street. (Courtesy of Thomas Walsh.)

EDISON VOLUNTEER FIRE COMPANY NO. 1

EX - CHIEFS

Theodore Fauquier	1924-29	Mitchell Erceg	1958
Rudolph Peins	1930	Arnold Hergenhan	1959
Leonard McLane	1931-32	Herman Good	1960
Joseph Stahl	1933-34	Joseph Haggerty	1961
Albert Eddy	1935	Charles Kohlbusch	1962
John Wilkens	1935-36	Joseph Haggerty	1963
Henry Koerber	1937	John Stanko	1964
Stewart Straka	1938	Donald Yackel	1965-66
Kenneth Shepard	1939	Charles Kohlbusch	1967
Andrew Dudas	1940-41	Bernard Ruthberg	1968
Stewart Straka	1942-45	Charles Kohlbusch	1969-70
Wesley Wilkens	1946	Morris Kaplan	1971
John Lambly	1947	John Guilfoyle	1972
Stanley Kebel	1948	Morris Kaplan	1973-75
Martin Dige	1949-52	Victor D'Addario	1976-77
Lee Sofield	1953	John Guilfoyle	1978
Robert Gray	1954	Walter Boychick	1979-88
Earl Evans	1955	Richard P. Campbell	1989-90
Walter Konesky	1956	Walter Boychick	1991-07
Willard Evans	1957		

Listed above are the chiefs that have led Edison Volunteer Fire Company No. 1 throughout the years.

Three

Raritan Engine Company No. 2

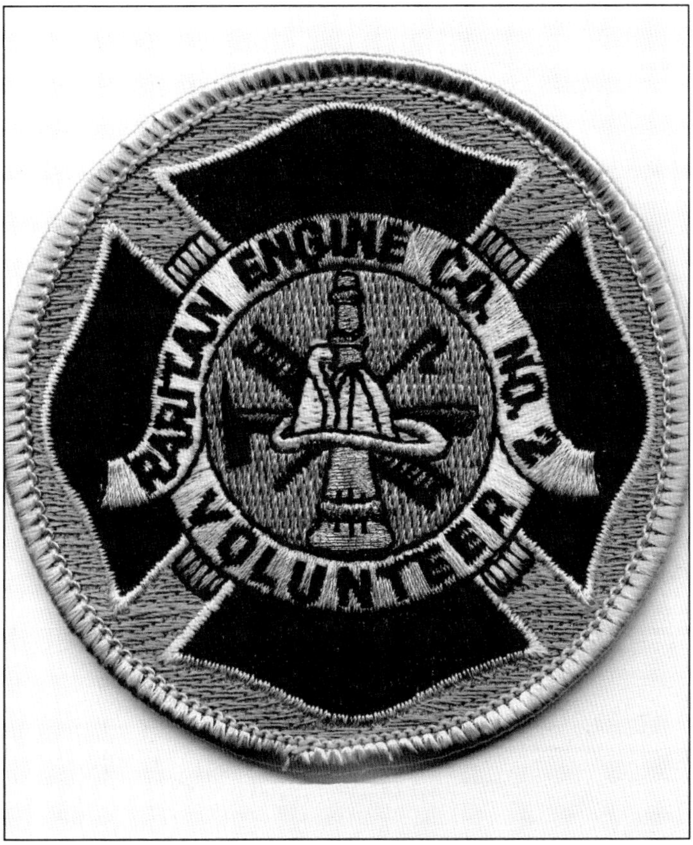

Raritan Engine Company No. 2 was formed in 1923 to protect the Clara Barton section of Raritan Township. This area, known as Fire District No. 3, encompassed the section between the Metuchen border on the west and the Woodbridge border on the east. In the 1960s, the company increased their response area to include the former site of the Raritan Arsenal, which has, throughout the years, developed into the largest industrial park on the east coast. Today the company protects the same area, now known as Precinct 3 and is the only Edison volunteer fire company that still remains in their original firehouse.

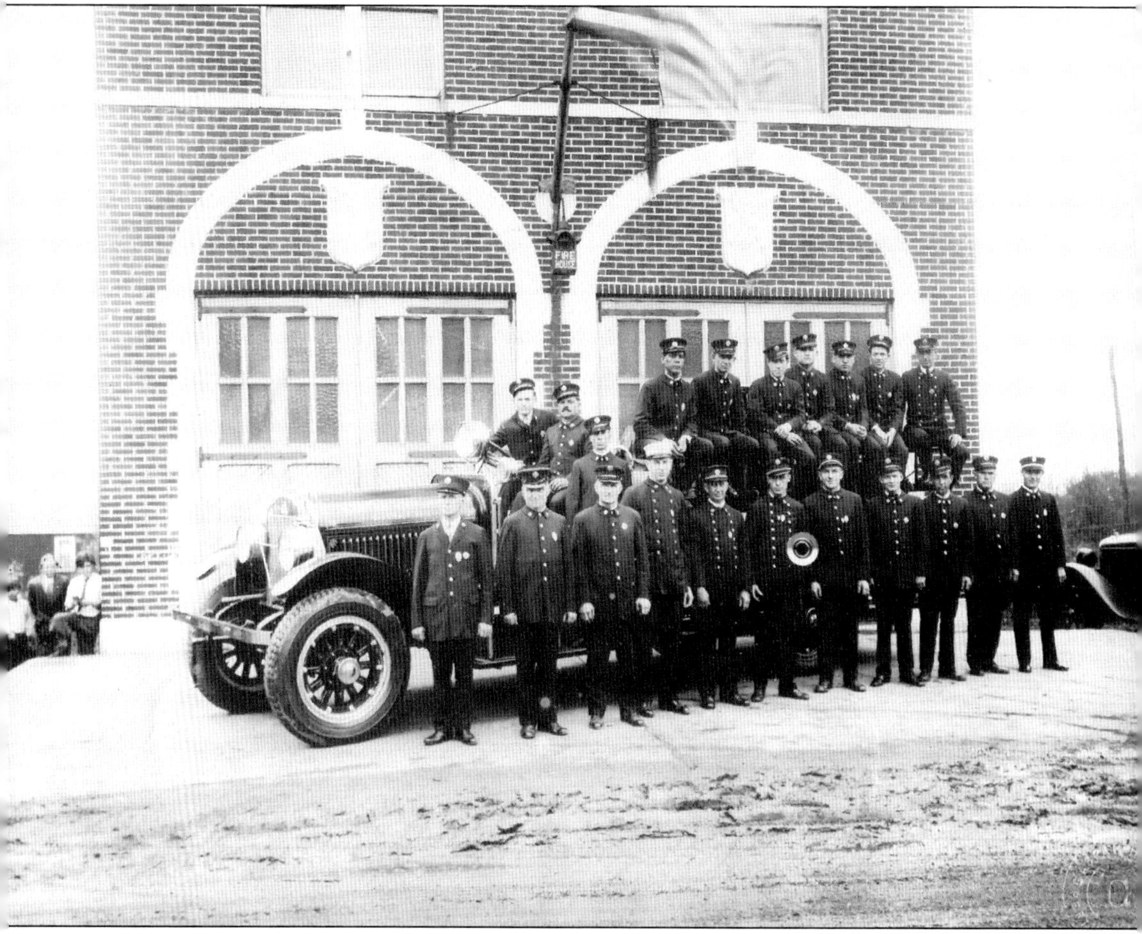

On August 17, 1923, Raritan Engine Company No. 2 received its charter and started providing fire protection in the Clara Barton section. The members elected Clifford Pfeiffer as the company's first fire chief. On July 1, 1924, William Gross, father of Fire Chief Nathan Gross (pictured above wearing the white cap), generously donated land on the corner of New Brunswick Avenue (now known as Amboy Avenue) and Charles Street (now known as Burchard Street) to the commissioners of Fire District No. 3 with the stipulation that the property would be used for fire department services only. By 1925, the fire commissioners purchased a 1924 Studz Bearcat fire engine, which was temporarily housed in one of the bays at Gross' Oil Company on the corner of Woodbridge Avenue and New Brunswick Avenue until the firehouse was built. Within a year, a beautiful two-story brick firehouse was erected. This photograph, which dates back to 1926, shows the original members of Raritan Engine Company No. 2 posing with their Studz Bearcat in front of their new firehouse. (Courtesy of Raritan Engine Company No. 2.)

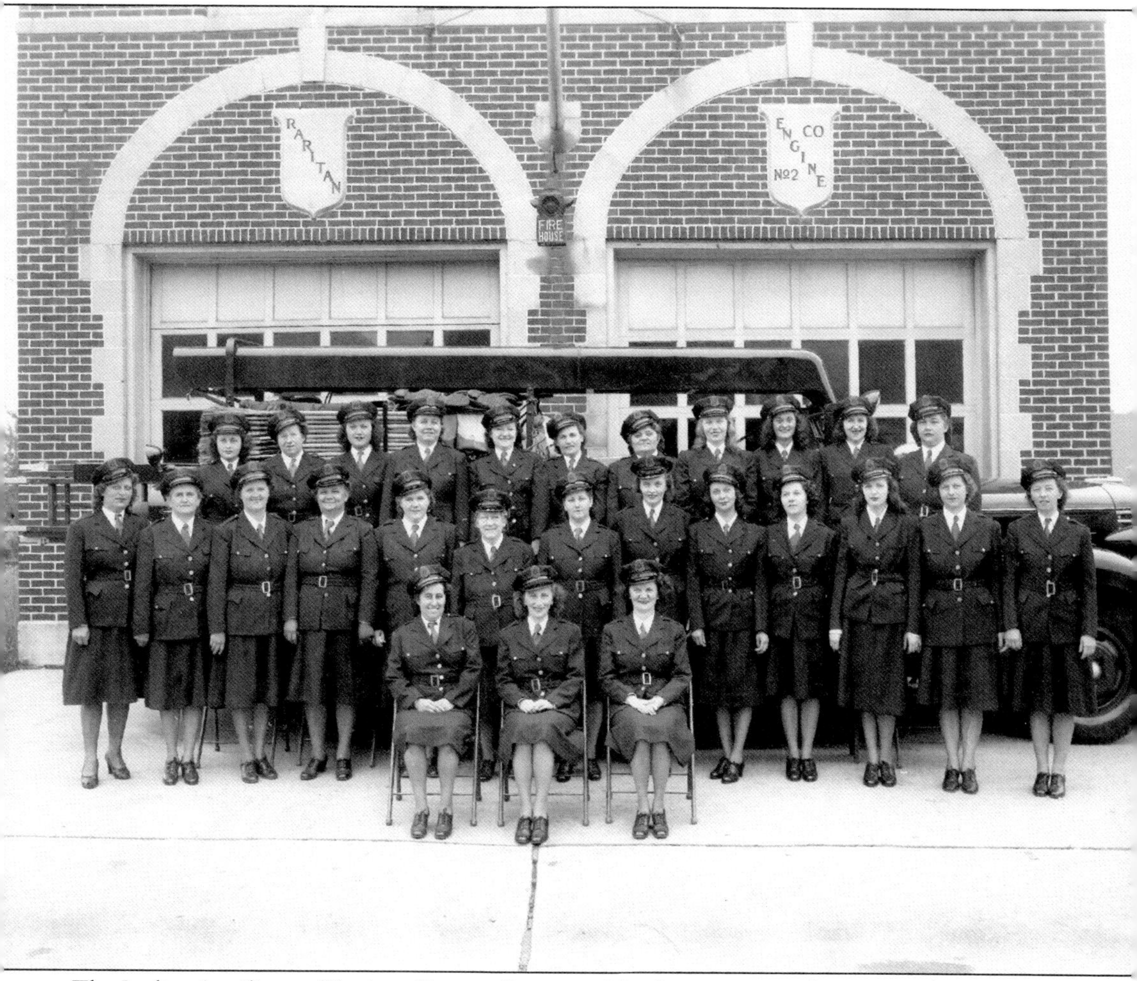

The Ladies Auxiliary of Raritan Engine Company No. 2 was organized in 1927. The group not only supported firemen with food and refreshments at fire scenes, but also organized dances and special events for the families and the community. This beautiful photograph was taken shortly after the ladies purchased their uniforms in the mid-1940s. Pictured from left to right are (first row) Margaret Kovacs, Adelia Kish, and Anna Dudics; (second row) ? Horniack, Anna Dudics, Mary Dudash, Mary Demcsak, Helen Madger, Elizabeth Larson, Catherine Nemeth, Julie Kearstan, Helen Demcsak, Mary Onder, Teresa Asprocolas, Margaret Nagy, and Josephine Kearstan; (third row) Dorothy Kovacs, Catherine Milcsik, Anna Lucas, Julia Lako, Mary Mulnar, Anna Kosup, Margaret Kalman, unidentified, Elizabeth Bednarik, ? Sabo, and Anna Demcsak. (Courtesy of Donald Dudics.)

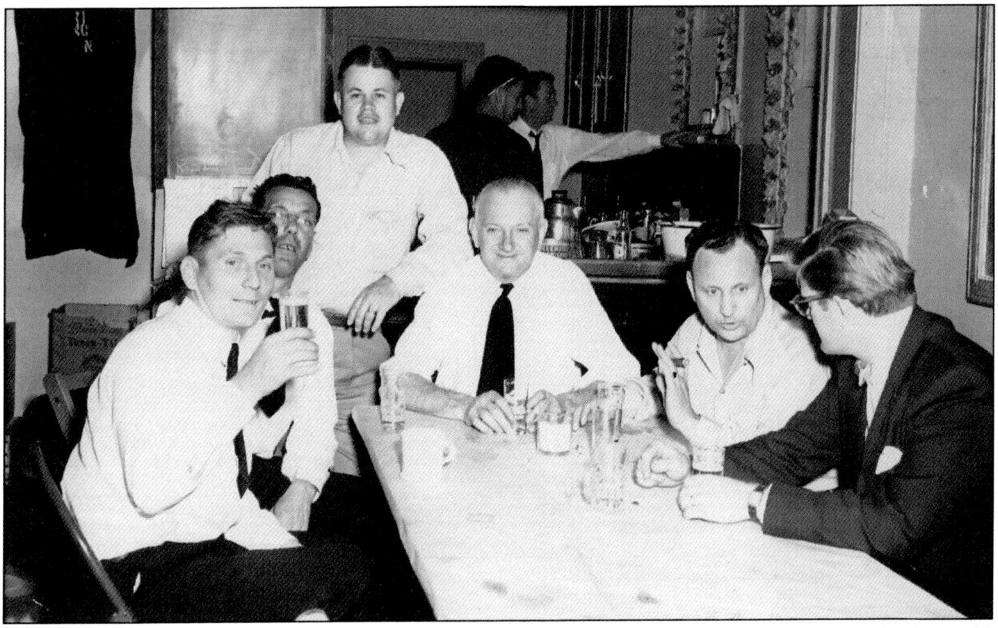

After marching in the Memorial Day parade, members of Raritan Engine Company No. 2 would return to the firehouse for lunch and refreshments. The firefighters would park the fire engines outside in order to use the apparatus floor for special occasions such as this. At the table, from left to right, are firefighters Joseph Sovart, John Vincz, Albert Antonides, Chief John Nagy, Peter Lucas, and Edward Soden. Beyond the access window are Milton Asprocolas and Michael Kearstan in the kitchen. (Courtesy of Joan Coleman.)

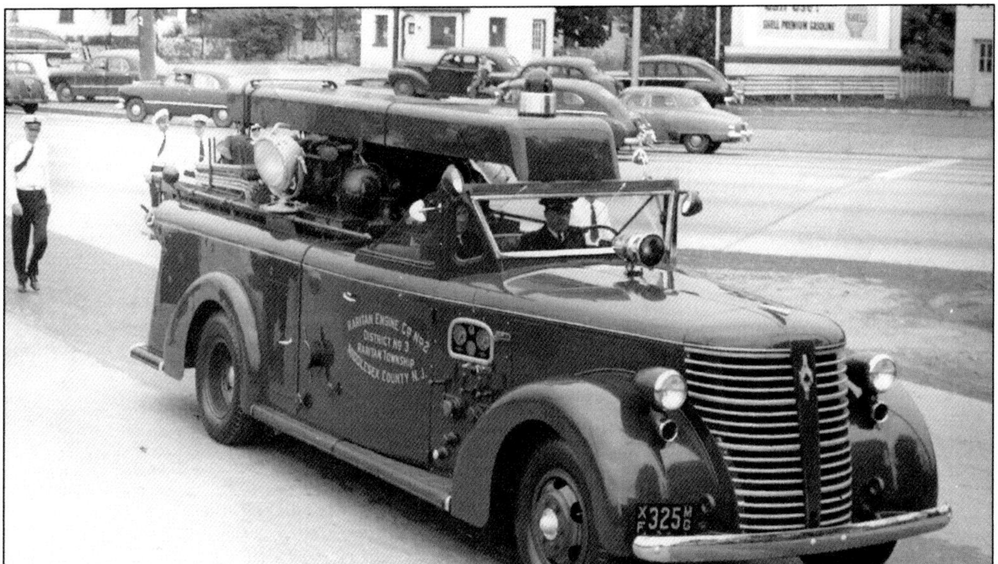

This photograph was taken at the intersection of Route 1 and Woodbridge Avenue during the 1955 Memorial Day parade where the members of Raritan Engine Company No. 2 are marching alongside their 1942 American LaFrance 500 series fire engine. This engine was unique due to its center-mounted ladder rack and small booster tank that carried only 100 gallons of water. (Courtesy of Raritan Engine Company No. 2.)

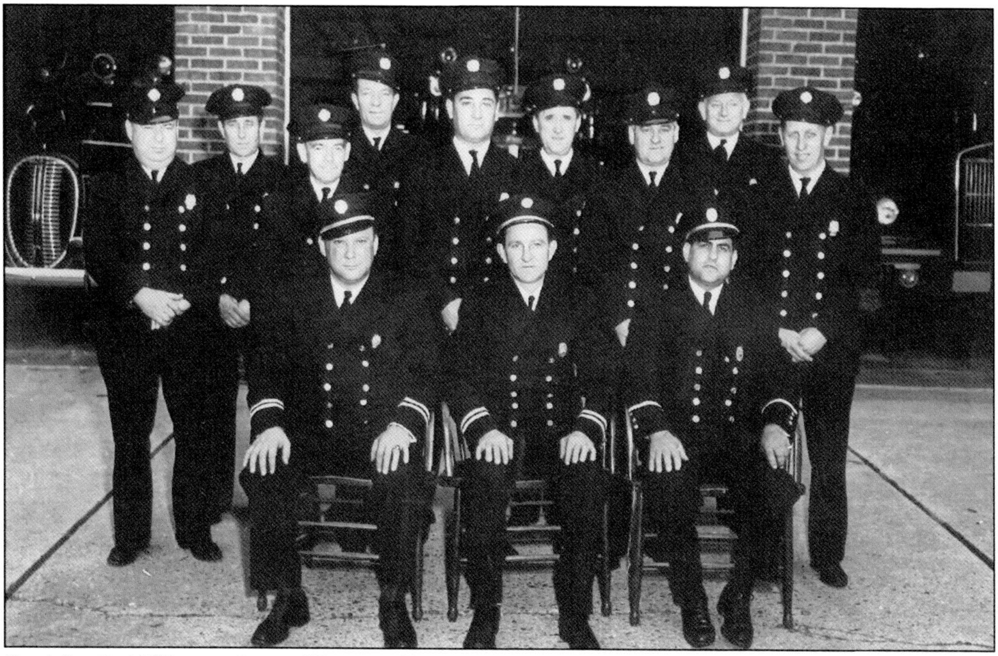

On May 2, 1955, members of Raritan Engine Company No. 2 pose in front of the firehouse on Plainfield Avenue after attending Raritan Engine Company No. 1's annual Fireman's Memorial Church Service in Fire District No. 1. Pictured from left to right are (first row) Steven Madger, Wilbert Blanchard, and Samuel Verella; (second row) Steven Lako, John Onder, John Lako, Michael Kearstan, Milton Asprocolas, Emery Kindle, Michael Dudash, John Nagy, and Louis Larson. (Courtesy of Joan Coleman.)

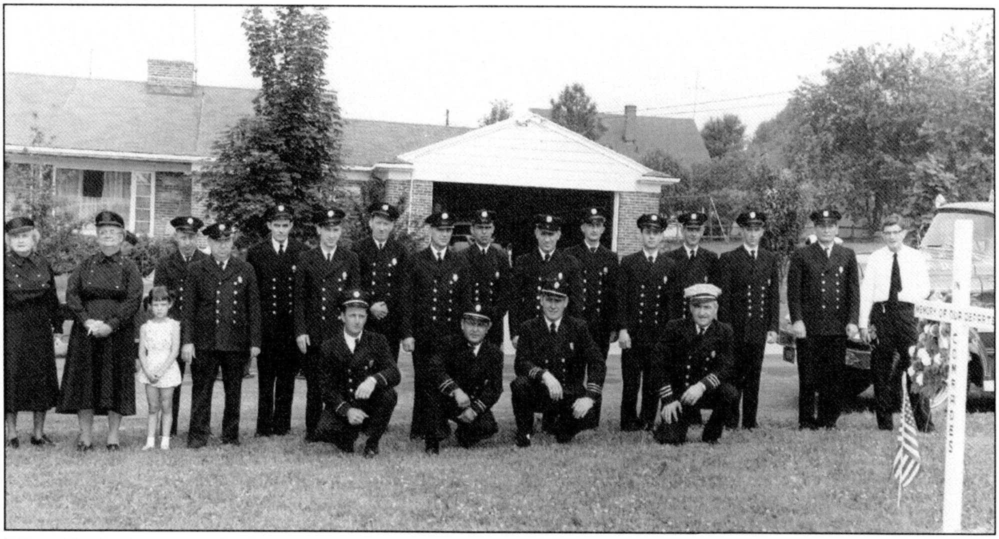

This 1959 photograph shows the members of Raritan Engine Company No. 2 gathered together for a moment of silence in front of the company's monument on Memorial Day. During the 1940s, members built a wooden cross on the corner of Woodbridge Avenue and Amboy Avenue to commemorate their departed members. In 1960, the company replaced the wooden cross with a large concrete monument, and they share the memorial with Clara Barton Post 324 American Legion. (Courtesy of Douglas Kosup.)

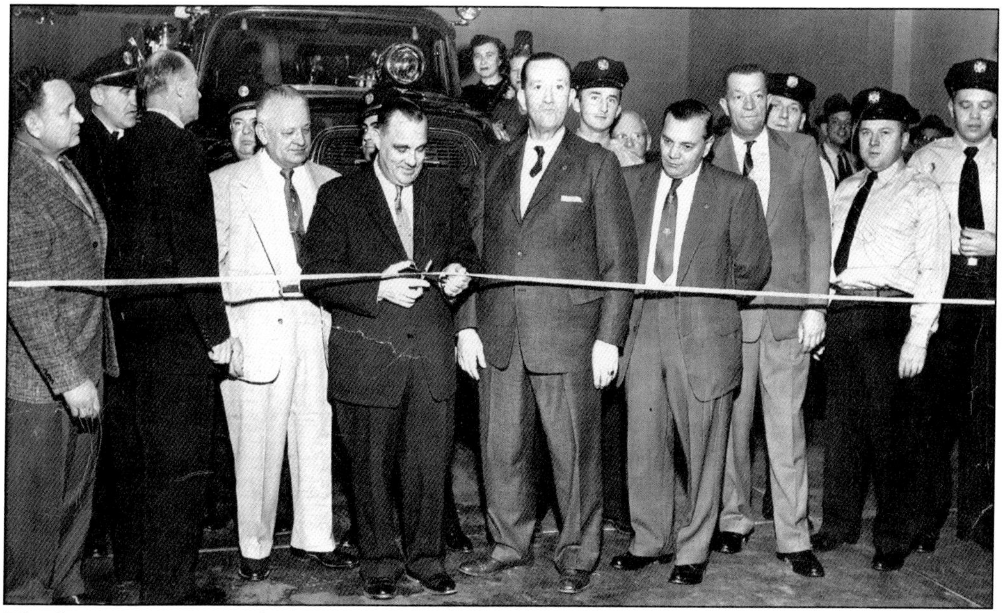

The fire commissioners of District No. 3 received approval from the public in 1955 to construct an addition onto their firehouse. The addition gave Raritan Engine Company No. 2 space to house two more apparatus with an alarm room in the rear at a cost of $80,000. The photograph above was taken in 1956 during the dedication ceremony with Mayor Thomas Swales cutting the ribbon. Pictured from left to right are Peter Lucas, Albert Kosup, commissioner Stephen Madger, Steven Lako behind commissioner Joseph Simon, Assistant Chief Samuel Verella behind Mayor Thomas Swales, public safety commissioner Julius Engel, Donald Dudics, John Vinsz, Michael Kearstan, Joseph Sovart, John Onder, and George Zigre. (Courtesy of Raritan Engine Company No. 2.)

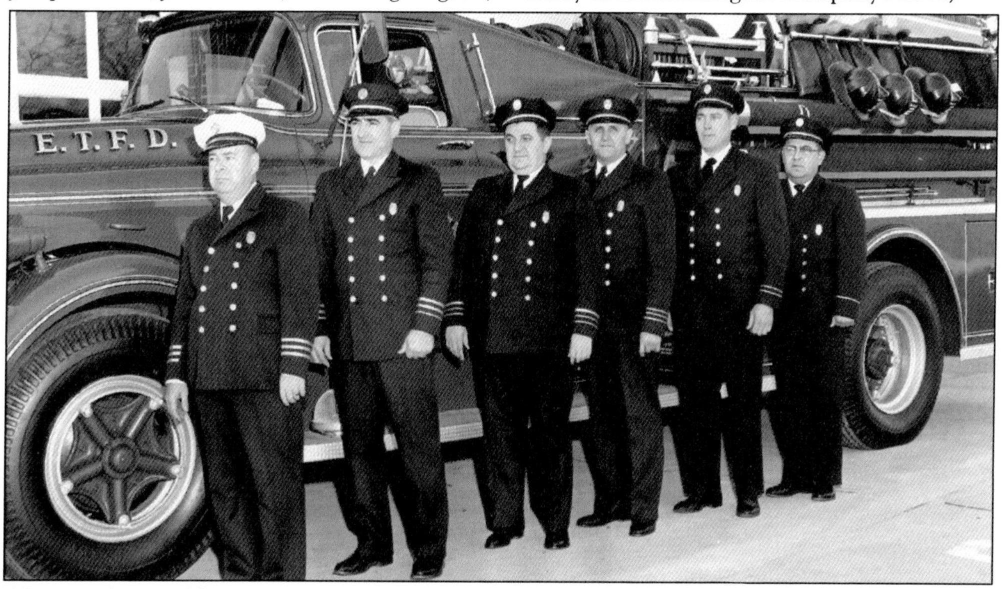

Along with an addition to Raritan Engine Company No. 2's firehouse, a new fire engine was purchased and dedicated on the same day in 1956. Proudly standing next to their new engine are (from left to right) Steven Lako, Albert Kosup, Louis Pulasty, Charles Nemeth, George Asprocolas, and Julius Yuhas. (Courtesy of Raritan Engine Company No. 2.)

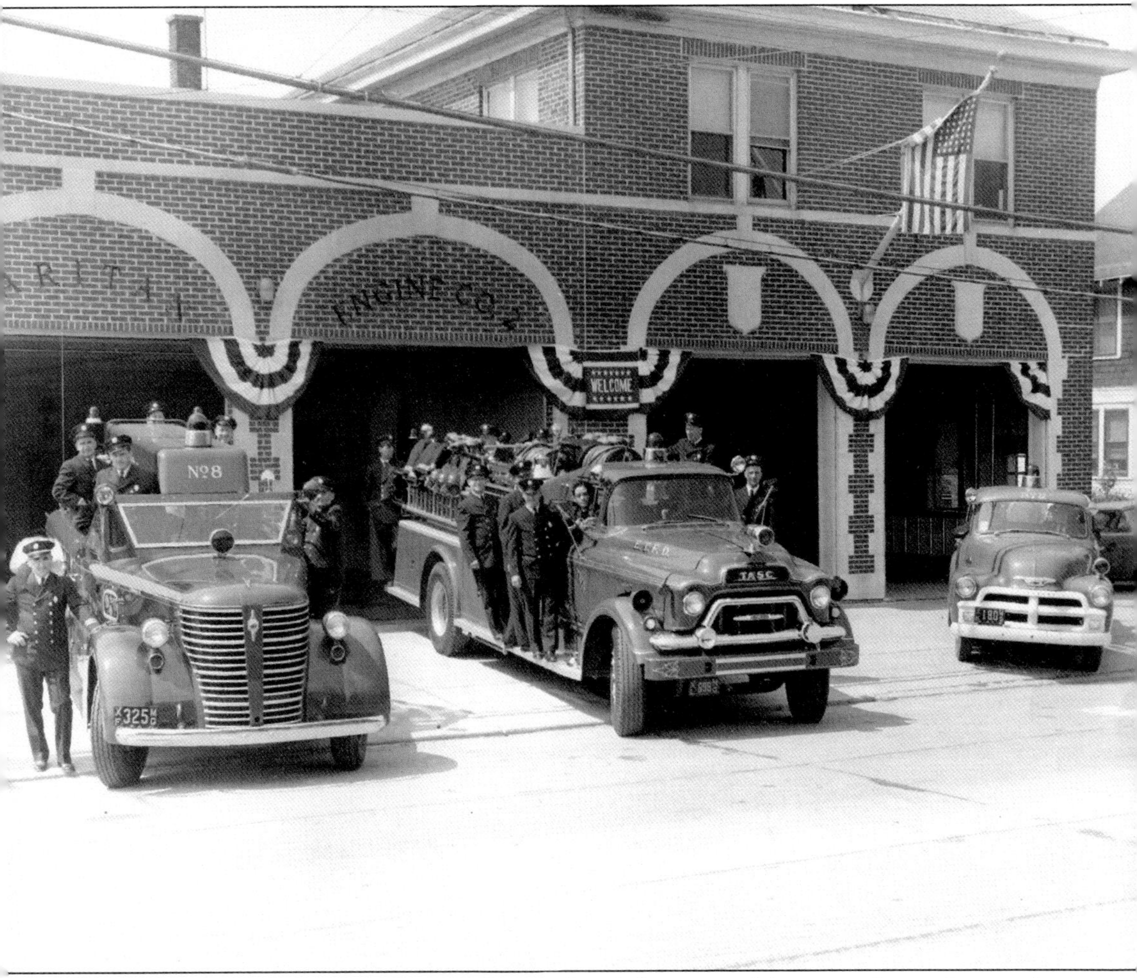

This photograph, which was taken during the dedication celebration, shows the two-bay expansion of the Raritan Engine Company No. 2 firehouse along with the members posing on their apparatus. The new GMC TASC is positioned in the center between the 1942 American LaFrance and the 1953 Chevrolet pick-up utility truck. A few years after this addition, the Division of Fire was established and this firehouse became Fire Headquarters. A small communications center was situated in the main office for career firefighters to dispatch fire and first-aid calls throughout Edison Township. (Courtesy of Douglas Kosup.)

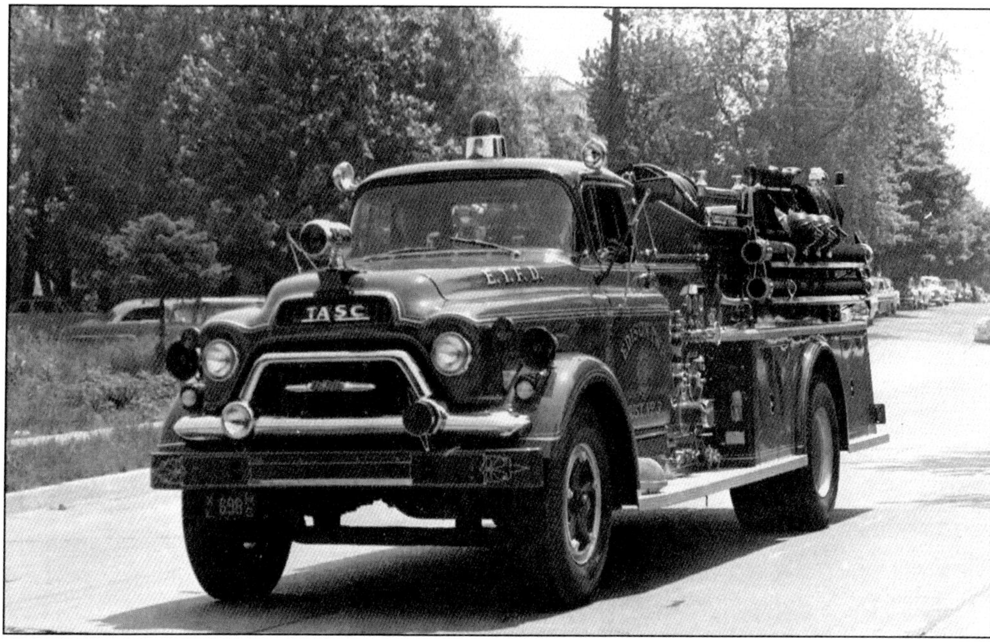

Fire District No. 3 paid almost $23,000 for this fully-equipped GMC that was built by Trautwein and Sons Company (TASC) in Woodbridge, New Jersey. It carried 300 gallons of water and was equipped with a pump capable of discharging 750 gallons of water per minute. (Courtesy of Douglas Kosup.)

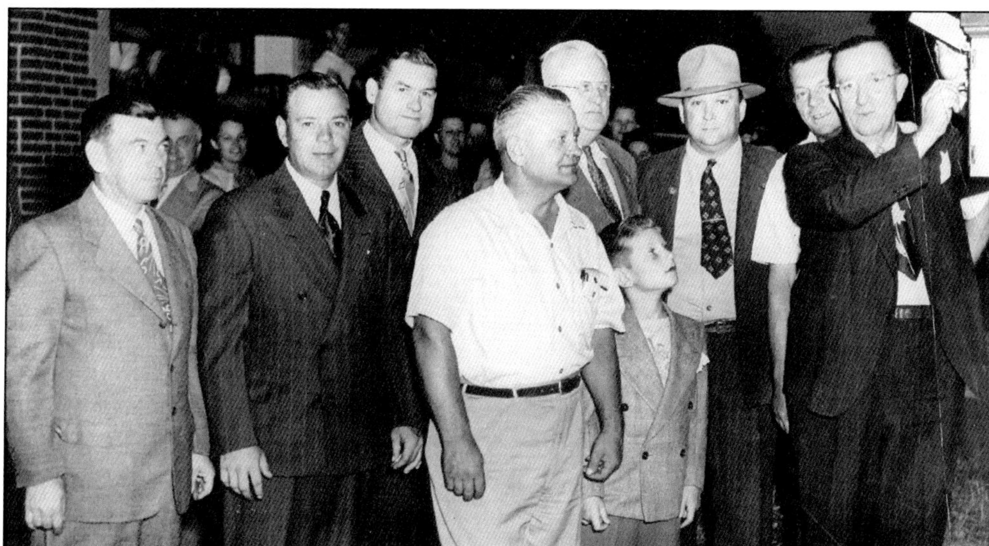

The fire commissioners of District No. 3 had several fire alarm boxes installed throughout the district. The alarm boxes would not only activate the community siren, but would also send a signal via telegraph wire to the firehouse identifying the box number. Shown above is the fire alarm pull box dedication ceremony in 1957. Pictured from left to right are John Lako, John Vincz, John "Duke" Dudash, Joseph Simon, unidentified, Edward Madger, Stephen Madger, and Michael Kearstan. Public safety commissioner Julius Engel is pulling the lever to the pull box mounted in front of the firehouse. (Courtesy of Donald Dudics.)

Fire District No. 3 was the only district in Edison Township to have fire alarm pull boxes. Alarm box No. 424, shown here, was mounted inside Roosevelt Hospital near the boiler room. After pulling the handle down, a signal was sent to the firehouse to notify the dispatcher of the alarm location. A total of 58 pull boxes remained in service until Chief H. Ray Vliet ordered for their removal by the end of 1979. (Courtesy of James Hook, photograph by Karen Daugherty.)

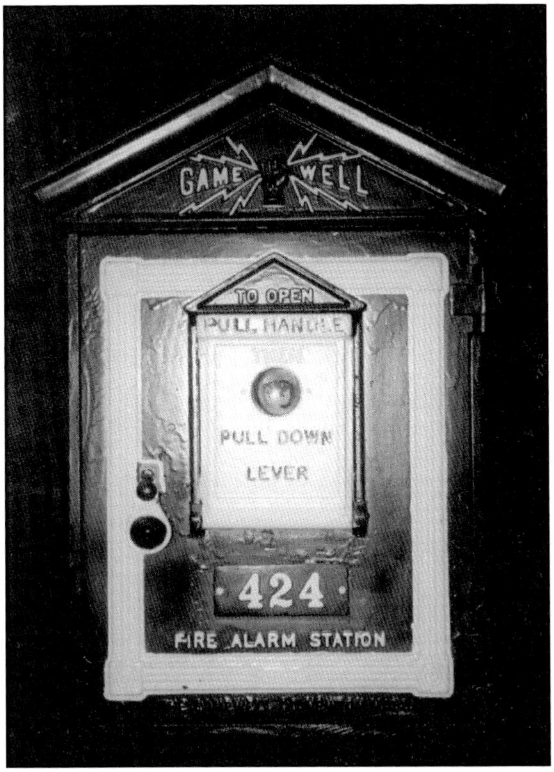

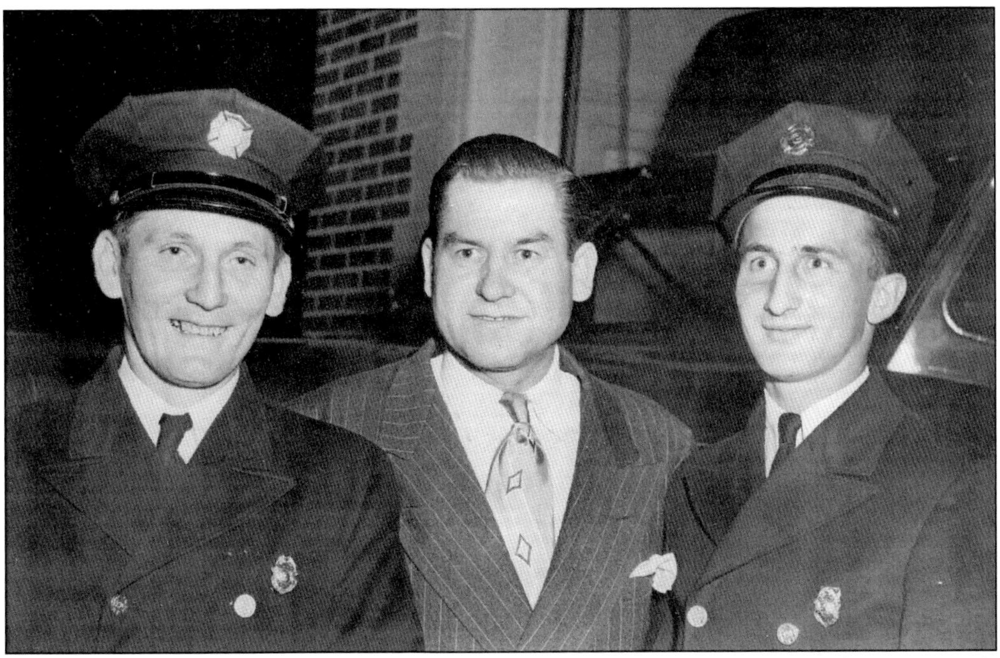

During this fire alarm pull box dedication, Fire Commissioner John "Duke" Dudash poses with firefighter Joseph Sovart (left) and firefighter Donald Dudics (right). Both Sovart and Dudics were hired in 1949, therefore becoming Fire District No. 3's first career firefighters. (Courtesy of Donald Dudics.)

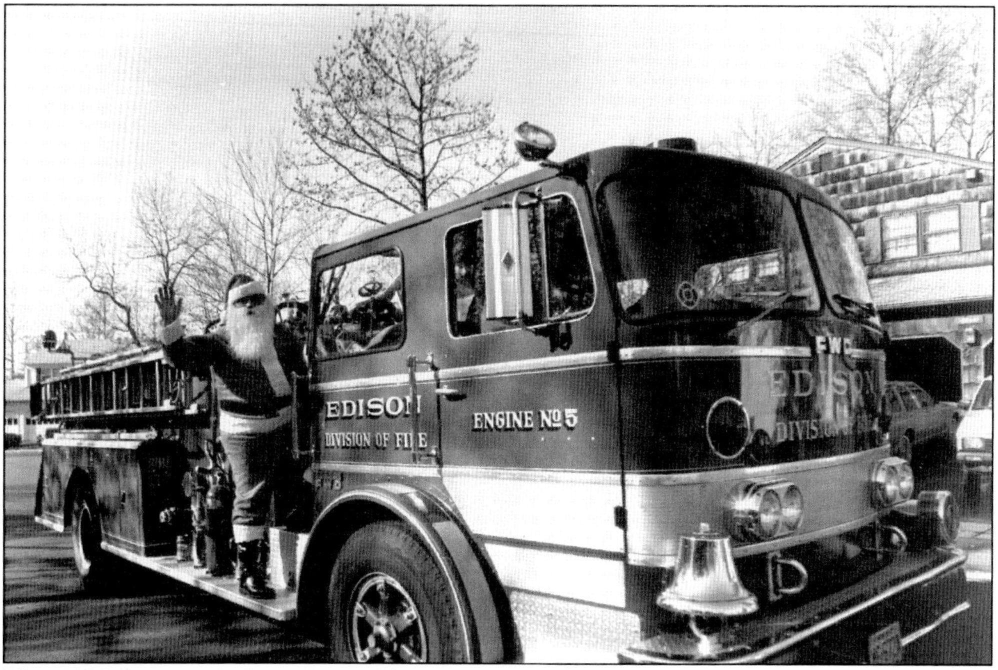

This 1986 photograph shows Santa Claus waving to Clara Barton residents while holding on to the 1966 FWD fire engine. The tradition of members of Raritan Engine Company No. 2 driving Santa throughout the precinct to distribute candy canes continues to be an exciting event for neighborhood children. (Courtesy of Jim D'Amico.)

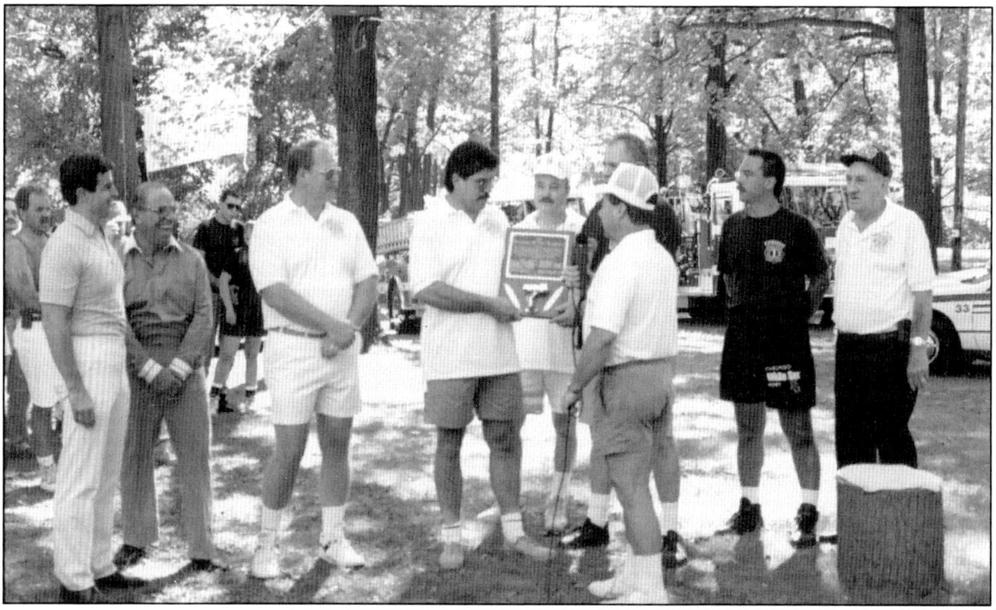

In 1993, both Raritan Engine Company No. 2 and Edison Volunteer Fire Company No. 1 received identical Pierce fire engines. During the dedication ceremony held at Roosevelt Park, council member David Papi, on behalf of the town council, received a special plaque from the fire chiefs to show their appreciation for the purchase of the apparatus. (Courtesy of Raritan Engine Company No. 2.)

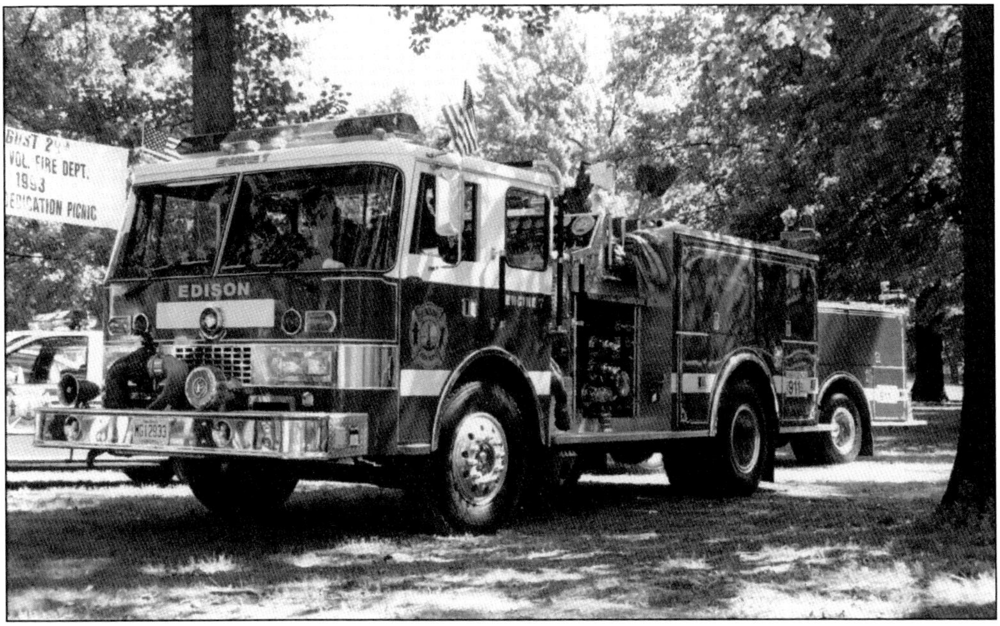

This pre-owned 1983 Pierce fire engine was purchased for Raritan Engine Company No. 2 because it has a mid-mounted pump that would place the pump operator in a safer position, especially during highway operations on the New Jersey Turnpike or Interstate 287, for example. (Courtesy of Raritan Engine Company No. 2.)

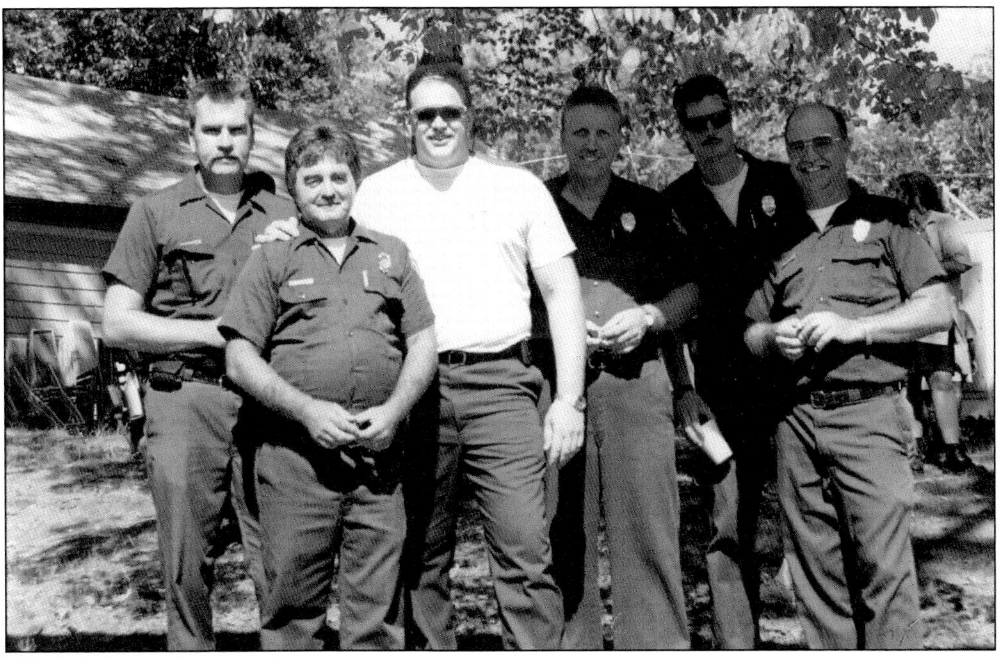

The members of Raritan Engine Company No. 2 invited career firefighters to join them at their dedication celebration in Roosevelt Park. Volunteer firefighter Robert Kearstan (in the white shirt) is gathered together with his career firefighter buddies (from left to right) Edward Koehler, Emery Tibok, Jack Campbell, Glenn Hunter, and Allan Yourstone Sr. (Courtesy of Raritan Engine Company No. 2.)

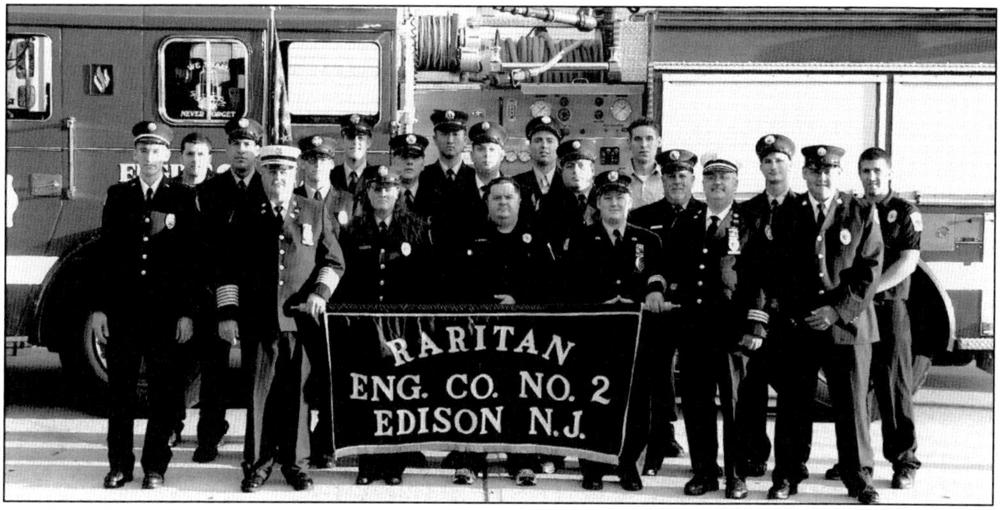

On June 11, 2007, the members of Raritan Engine Company No. 2 gathered together to take a formal photograph in front of the 1997 Seagrave fire engine. Pictured from left to right are (first row) Anton Getz, Chief Robert Barrett, Kimberly Barrett, Jeff Davids, Paul Nannery, Assistant Chief James Nannery, and David Ho; (second row) Adelis Arrevillagas, Michael Weigel, Robert Silagy, Christopher Latham, Louis Silva, Gary Guzzi, Matthew Gabriel, and Timothy Kapitan; (third row) Christopher DeMatteo, Brian Jones, Dong Hee Lee, Andrew Latham, and Timothy Asprocolas.

RARITAN ENGINE COMPANY NO. 2

EX - CHIEFS

Clifford Pfieffer	1923-25
Albert Antonides	1925-26
Nathan Gross	1926-27
Milton Gross	1927-28
Einer Jensen	1928-29
Louis Nagy	1929-30
Stephen Serenska	1930-31
John Kalman	1931-32
John Nagy	1932-33
James Asprocolas	1933-34
Joseph Simon	1934-35
Albert Kosup, Sr	1935-36
Michael Kearstan	1936-37
Michael Dudash	1937-38
Steven Simon	1938-39
Michael Bandics	1940
Steven Kurry	1941
Joseph Dudash	1942
John Kearstan	1943
Steven Jacob	1944
Julius Bartha	1945
John Dudics	1946
John Vincz	1947
Carl Benz	1948
John Dudash	1949
Peter Lucas	1950
Emery Kindle	1951
Michael Sovart	1952
John Onder	1953
Joseph Estok	1954
Wilbur Blanchard	1955
Steven Madger	1956
Samuel Verrella	1957
Steven Lako	1958
Albert Kosup, Jr	1959
Louis Pulasty	1960
Charles Nemeth	1961
George Asprocolas	1962
Christian Seich	1963
Frank Galya	1964
Edward Saranczak	1965
Robert Steinbach	1966
Edward Kraszewski	1967
John Soden	1968
William LaPointe	1969
Edwin Gadek	1970
Joseph Sovart	1971
Robert Kearstan	1972
Edward Madger	1973
Kenneth Kuzma	1974
Theodore Wolan	1975
David Fauquier	1976
Ronald Laurentiev	1977
Louis Wodash, Sr	1978
Jack Scully	1979-80
Richard Rosen	1981-84
Ronald Laurentiev	1985-86
Jack Scully	1987-01
Robert Barrett	2002-07

Listed above are the chiefs that have led Raritan Engine Company No. 2 throughout the years.

Four

H. K. Volunteer Fire Company

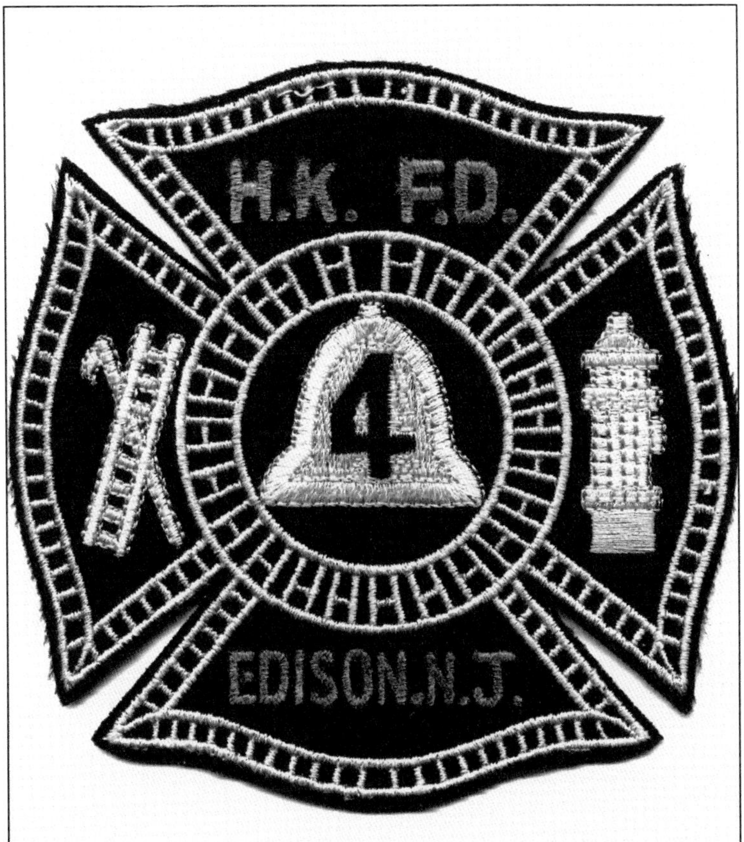

In 1926, the H. K. Volunteer Fire Company was formed to protect the citizens of northern Raritan Township, which encompassed the small section between Oak Tree Road, New Dover Road, Grove Avenue, and Wood Avenue. The company's response area increased in 1967 when the section between New Dover Road and the Clark border was added. This company patch was designed with the number four encircled by a bell and a border of bricklike stitching to symbolize H. K. Volunteer Fire Company's monument in front of the firehouse.

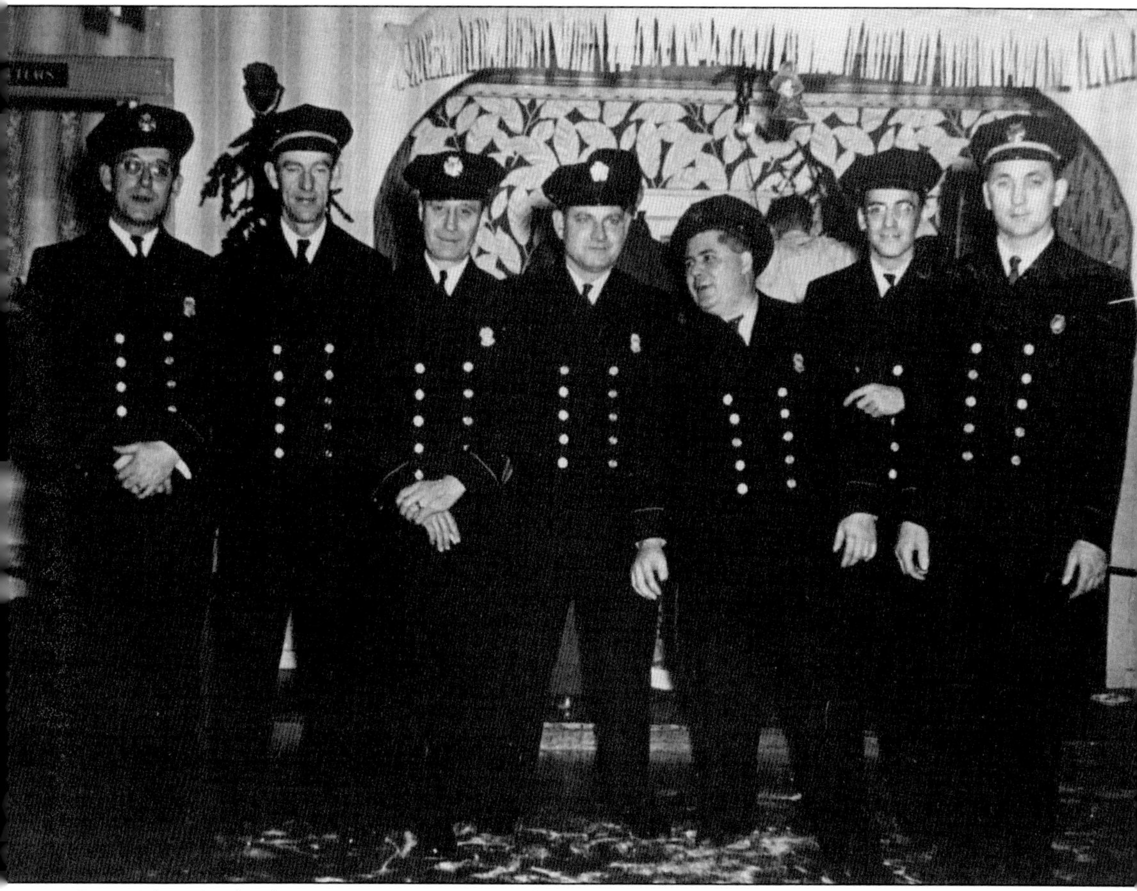

In 1925, a fire broke out in an apartment building located on the corner of Oak Tree Road and Henry Street. The fire destroyed the structure, which devastated local residents. Henry Kuntz, a local construction developer, helped the community by donating a lot on Henry Street for construction of a new firehouse. George Foster, a former resident of the recently destroyed apartment building, donated the lumber, and residents helped with the construction. Through the generous contributions of Kuntz and Foster as well as the dedication of a community to safeguard its citizens, the firehouse was completed and the company received its charter from Raritan Township on August 31, 1926. The fire company was named the H. K. Volunteer Fire Company in honor of Henry Kuntz. Shown above is the oldest membership photograph of the company's collection, which dates back to 1953. Pictured from left to right are Evans Lindquist, Harold Divitz, Walter Frank, Theodore Borowsky, Clarence Brunt, John Asanio, and Steven Burylo. (Courtesy of John Pencak.)

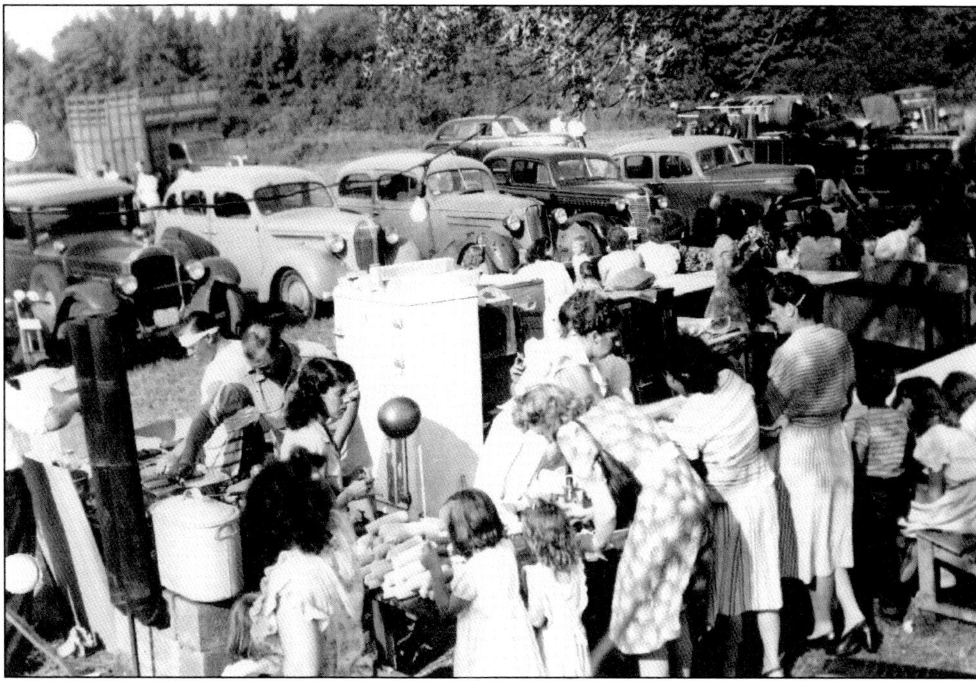

This 1948 photograph was taken during an annual H. K. Volunteer Fire Company picnic. The picnic took place in the fields off Henry Street. Notice H. K. Volunteer Fire Company's first fire engine, a 1922 Seagrave, located in the upper right corner of the photograph. The company purchased this previously owned fire engine in the late 1920s, and it remained in service for 25 years. (Courtesy of Patrick Novia.)

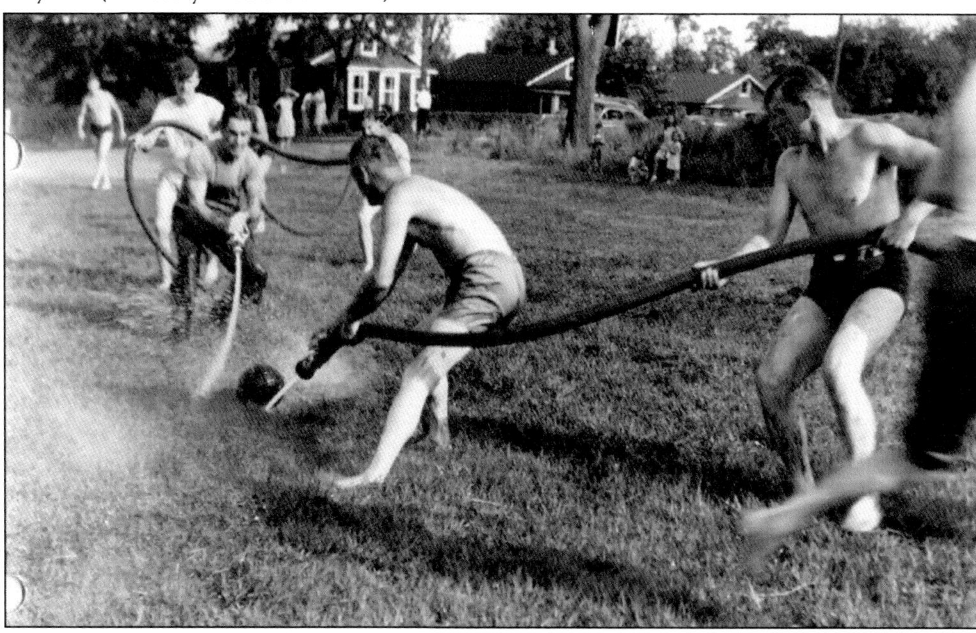

This 1948 photograph shows the members of H. K. Volunteer Fire Company playing a friendly game of "water polo" in the field. Instead of kicking the ball, they would use the water pressure from hose lines to try and push the ball into their opponent's goal. (Courtesy of Patrick Novia.)

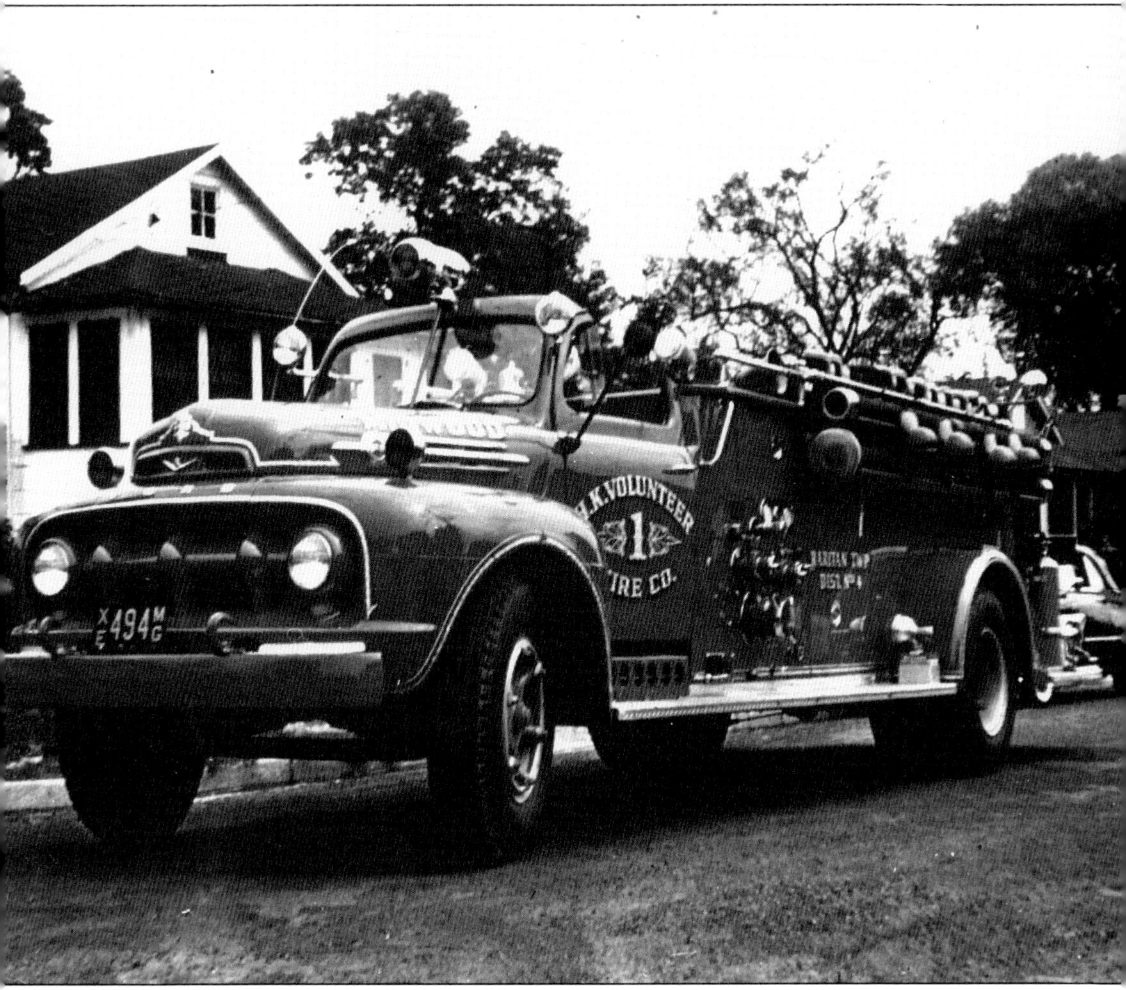

The commissioners of Fire District No. 4 decided that a new fire engine was needed because the original 1922 Seagrave was old and beyond repair. Therefore, in 1952, H. K. Volunteer Fire Company received a new 500-gallon-per-minute Ford fire engine. This photograph of the new engine was taken on Henry Street across from the firehouse. Notice the holes under the cab door, which were used to hold glass bulbs containing carbon tetrachloride, a chemical that was used as a suppression agent but is no longer used in the fire service due to its carcinogenicity and potential to generate lethal phosgene gas. Also, notice the name "Midwood" painted on the hood of the engine, which represented their small section of Raritan Township. (Courtesy of H. K. Volunteer Fire Company.)

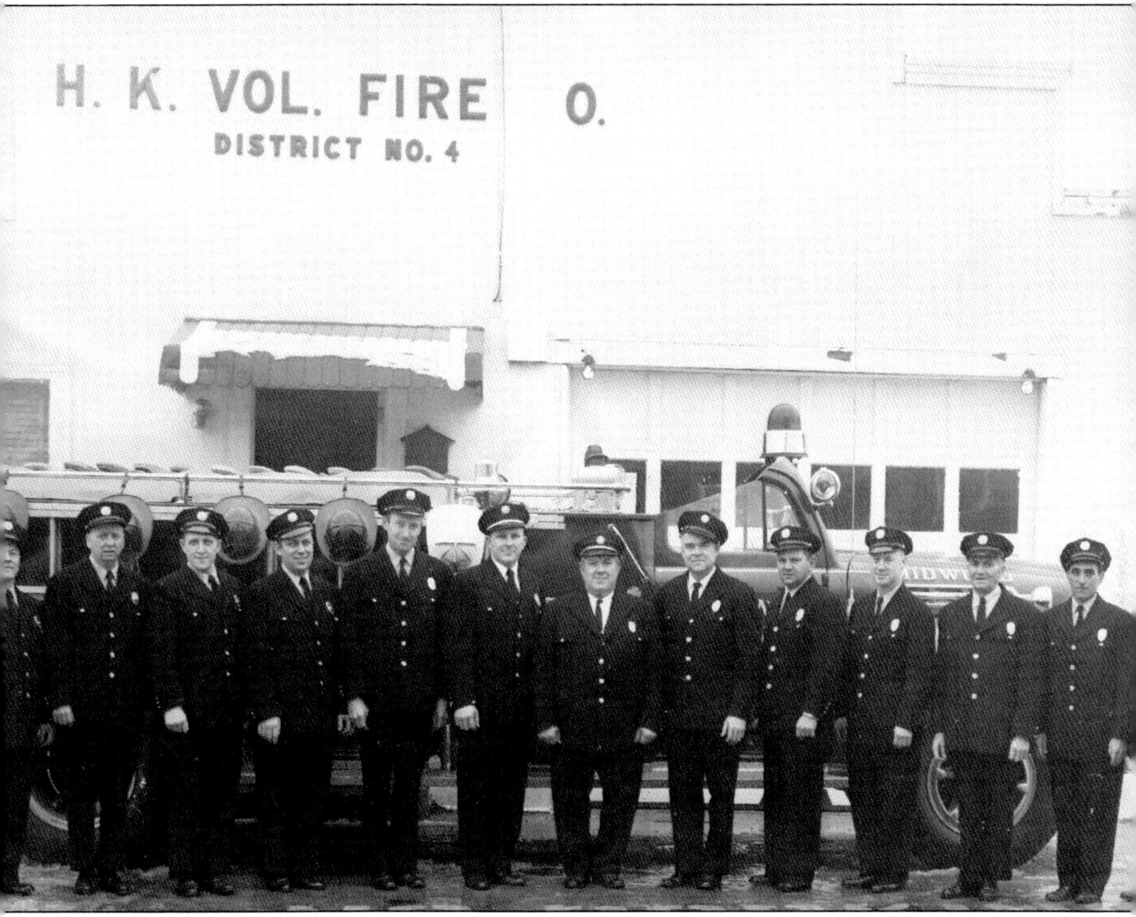

H. K. Volunteer Fire Company members pose in front of the Henry Street firehouse in 1959. Pictured from left to right are Joseph Jadus, Walter Karlo, Anthony Lichowid, Steven Mozsgae, John Pencak, Davis Cherick, Michael D'Allesandro, Charles Whitecavage, Tomas Kaczenski, Fredrick March, Frank Stanziola, Joseph Santos, and Ludwig Roell. (Courtesy of H. K. Volunteer Fire Company.)

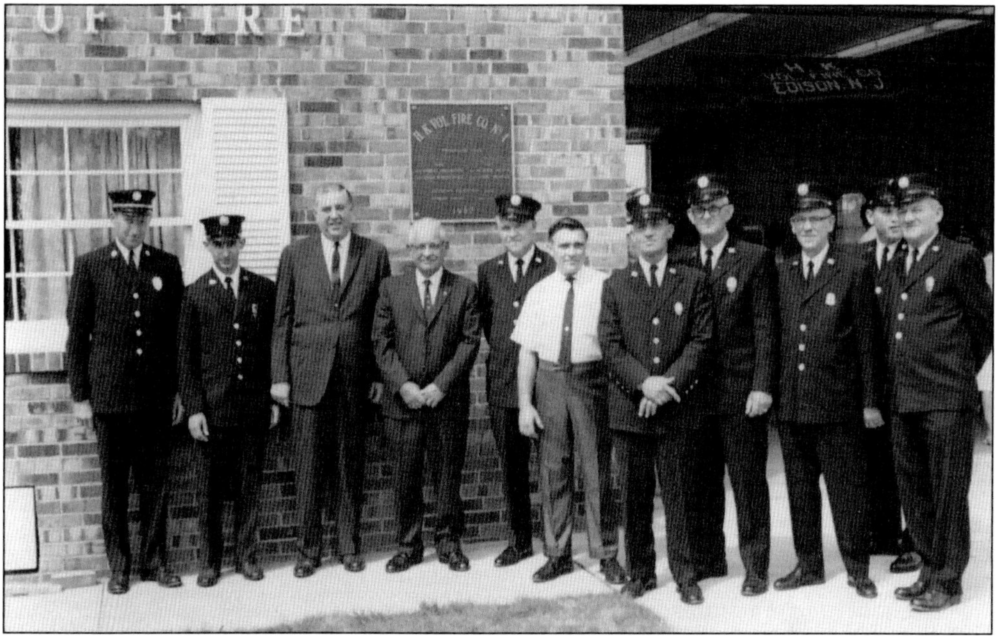

In the late 1950s, H. K. membership decided that the Henry Street firehouse was too small for their needs. By 1960, the board of education subdivided their lot on New Dover Road and a few years later, the new H. K. Volunteer Fire Company firehouse was built. Shown above is the dedication of the new firehouse on August 12, 1967. Pictured from left to right are H. K. volunteer chief John Pencak, president Patrick Novia, Mayor Anthony Yelencsics, Division of Fire supervisor Joseph Simon, Charles Whitecavage, Robert Wolent, Alphonse Novia, Joseph Masanet, Martin Lillis, Gerald Sarintino, and George Bierwirth. The photograph below shows the members officially "christening" the firehouse with a typical "wet-down." (Courtesy of H. K. Volunteer Fire Company.)

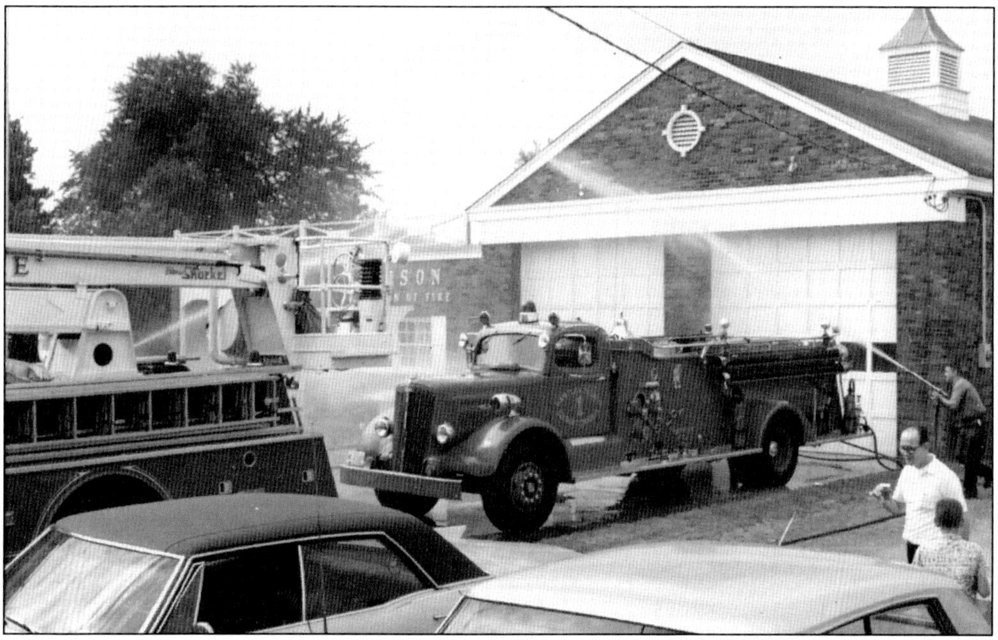

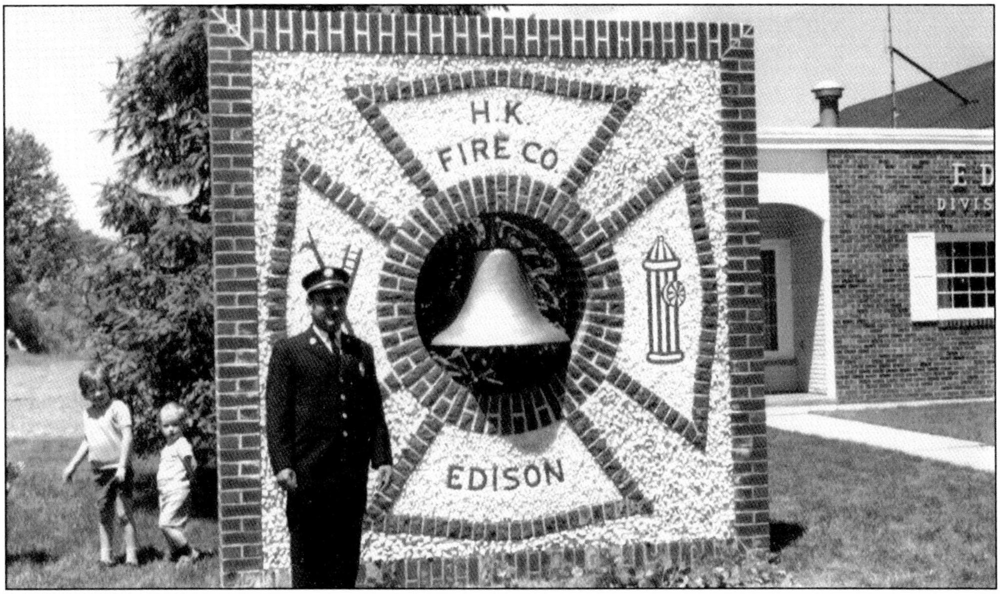

In 1971, firefighter Ludwig Roell decided to use his masonry skills to build a monument to honor firefighters. With the help of his masonry employee, Robert Kammlott, and volunteers from H. K. Volunteer Fire Company, Roell was able to build this large structure. The Maltese cross is outlined in traditional red brick that was chipped by hand to provide a rough appearance and the excess chips were used for the words and objects inside the cross. Roell proudly poses in front of the H. K. Volunteer Fire Company monument during the dedication ceremony on May 29, 1972. (Courtesy of Ludwig Roell.)

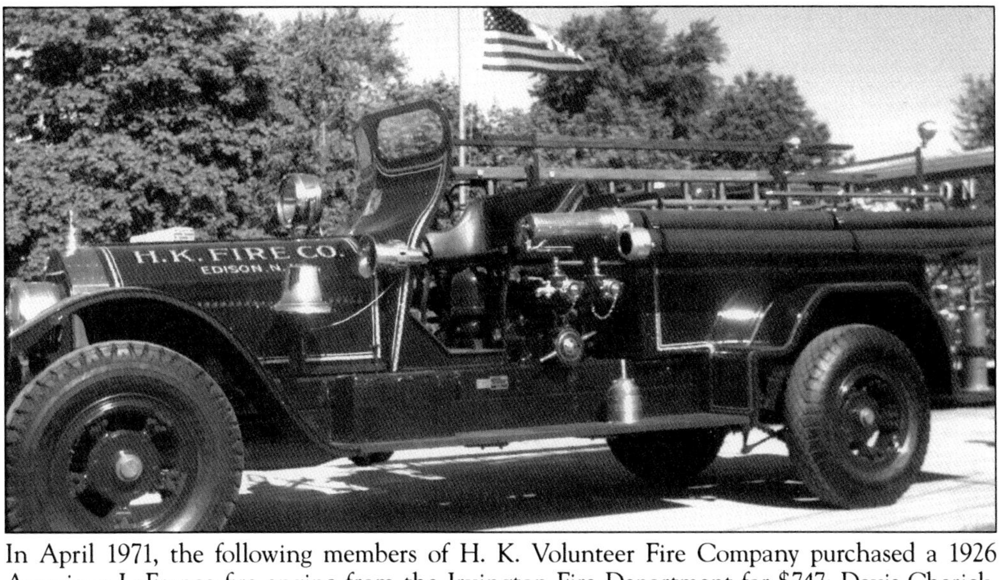

In April 1971, the following members of H. K. Volunteer Fire Company purchased a 1926 American LaFrance fire engine from the Irvington Fire Department for $747: Davis Cherick, Edward Koehler Sr., Thomas McKay, Patrick Novia, John Pencak, Ludwig Roell, Raymond Volker, Charles Whitecavage, Robert Wolent, and Louis Zavis. These 10 members spent over one year restoring the apparatus to the condition shown above. In fact, their antique has won several plaques and awards granted from antique fire shows throughout New Jersey. (Courtesy of Marion Cherick.)

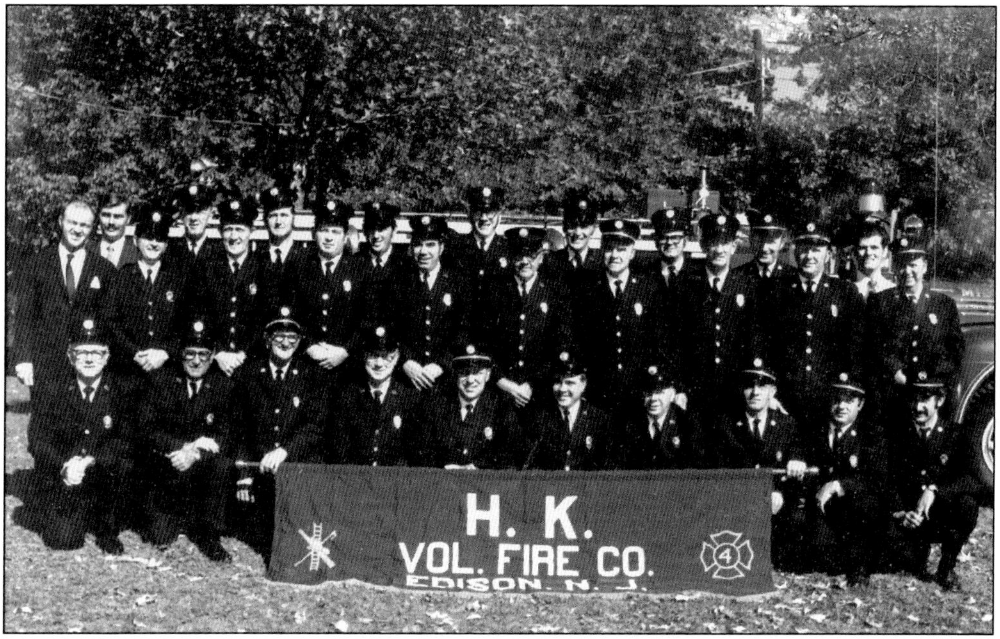

H. K. Volunteer Fire Company members pose outside in front of the 1952 Ford fire engine on Memorial Day 1972. The members from left to right are (first row) Joseph Jadus, Joseph Santos, William Fairbanks, Fredrick March, Steven Moszgae, Robert Wolent, Raymond Volker, Alphonse Novia, Ludwig Roell, and Patrick Novia; (second row) Eugene Enfield, Dennis Kubert, Robert Brunt, Thomas McKay, Anthony Lichowid, Robert Macyszyn, John Klamber, Joseph Sarintino, Rudolph Vivadelli, John Dinuzzo, Albert Kallai, Robert Hallbauer, Charles Whitecavage, Louis Zavis, Edward Koehler, John Pencak, Davis Cherick, Richard Hill, and Walter Karlo. (Courtesy of H. K. Volunteer Fire Company.)

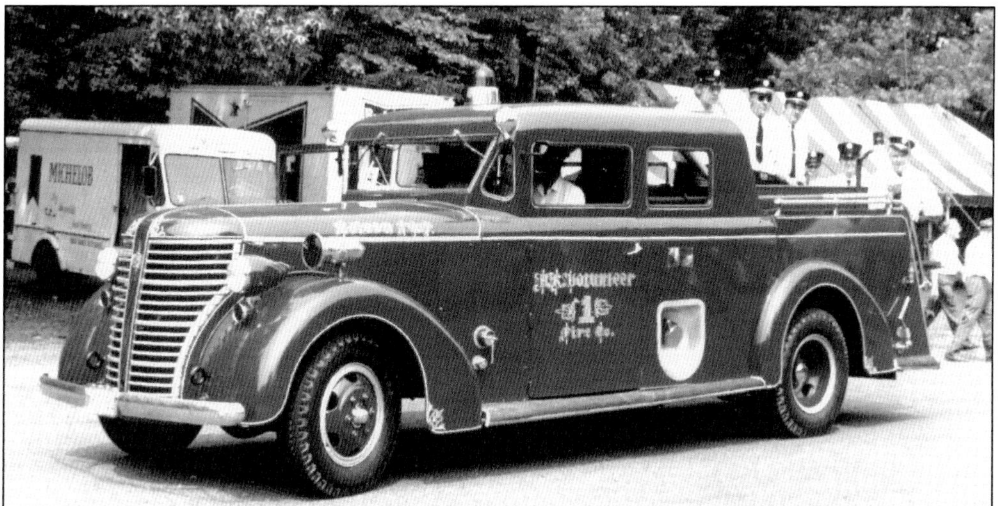
In 1965, H. K. Volunteer Fire Company purchased a 1941 American LaFrance fire engine from the Woodbridge Fire Company for $1. The apparatus had a wide and roomy cab, which required a firefighter with long legs to operate the four-speed manual transmission. It was used only for special events such as this 1974 celebration at Menlo Park's 50th anniversary. (Courtesy of Edison Volunteer Fire Company No. 1.)

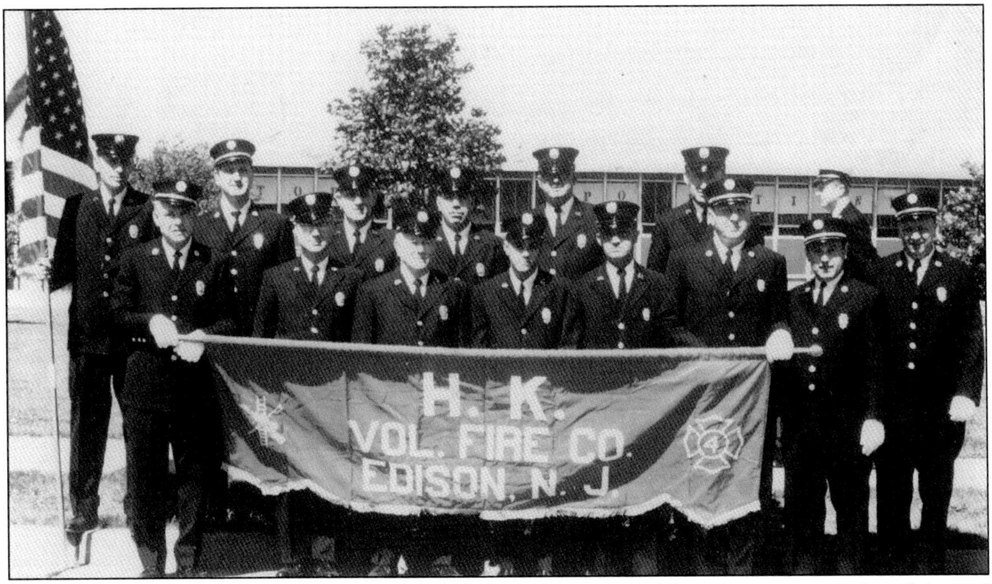

In 1976, H. K. Volunteer Fire Company and Oak Tree Volunteer Fire Company jointly celebrated their 50th anniversaries. A large parade, which was attended by famous actresses and former Edison residents Susan Surandon and Gail Fisher, ended with a big celebration at the James Madison Primary School parking lot. Photographed at the celebration from left to right are H. K. Volunteer Fire Company members (first row) Charles Whitecavage, Frederick March, Joseph Jadus, Richard DePaolo, Alphonse Novia, Davis Cherick, Louis Roell, and Walter Karlo; (second row) Robert Hallbauer, Robert Macyszyn, Anthony Lichowid, Rudolph Vivadelli, Louis Zavis, and John Pencak. (Courtesy of John Pencak.)

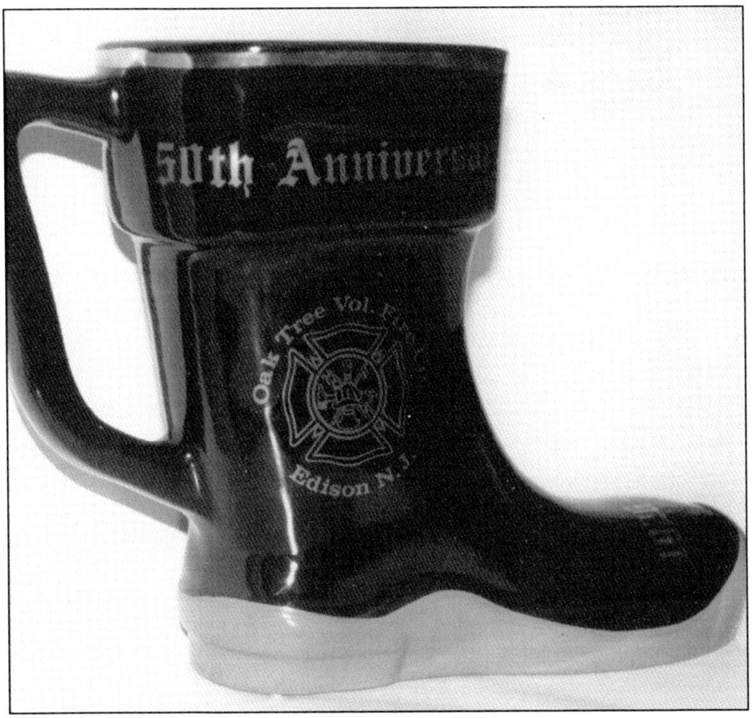

During an early planning meeting for the 50th anniversary celebration of H. K. and Oak Tree Volunteer Fire Companies, members conceived an innovative idea to raise money for the festivities. They revolutionized the fire souvenir business by creating a 40-ounce beer mug in the shape of a fire boot. The success of the H. K./Oak Tree 50th Anniversary Boot Mug exceeded all expectations. (Courtesy of Eugene Enfield Sr.)

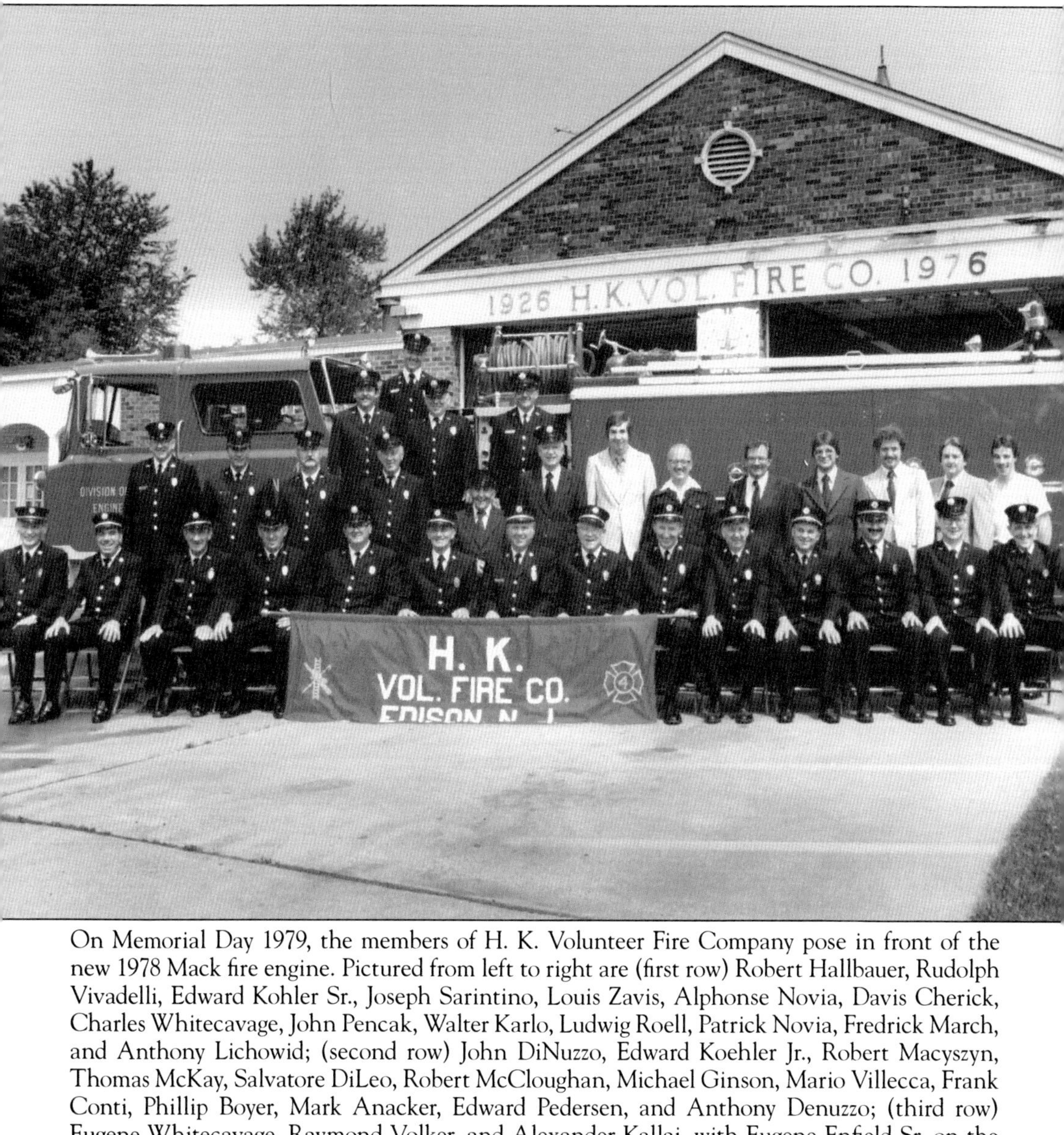

On Memorial Day 1979, the members of H. K. Volunteer Fire Company pose in front of the new 1978 Mack fire engine. Pictured from left to right are (first row) Robert Hallbauer, Rudolph Vivadelli, Edward Kohler Sr., Joseph Sarintino, Louis Zavis, Alphonse Novia, Davis Cherick, Charles Whitecavage, John Pencak, Walter Karlo, Ludwig Roell, Patrick Novia, Fredrick March, and Anthony Lichowid; (second row) John DiNuzzo, Edward Koehler Jr., Robert Macyszyn, Thomas McKay, Salvatore DiLeo, Robert McCloughan, Michael Ginson, Mario Villecca, Frank Conti, Phillip Boyer, Mark Anacker, Edward Pedersen, and Anthony Denuzzo; (third row) Eugene Whitecavage, Raymond Volker, and Alexander Kallai, with Eugene Enfield Sr. on the top. (Courtesy of Eugene Enfield Sr.)

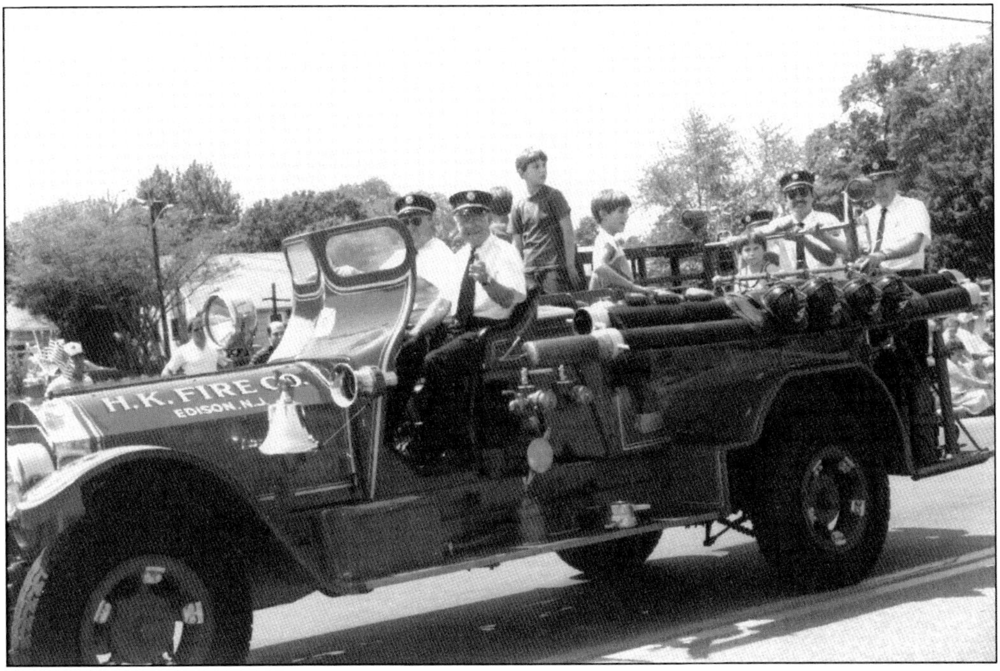

This 1985 photograph shows the members of H. K. Fire Company having fun during the Memorial Day parade. Davis Cherick is driving, Charles Whitecavage is pointing with a smile, and Ludwig Roell, Richard Buda, and John Pencak are standing on the back step. Youngsters, who included Patrick J. Novia and George Roell, are enjoying the ride as well. (Courtesy of Eugene Whitecavage.)

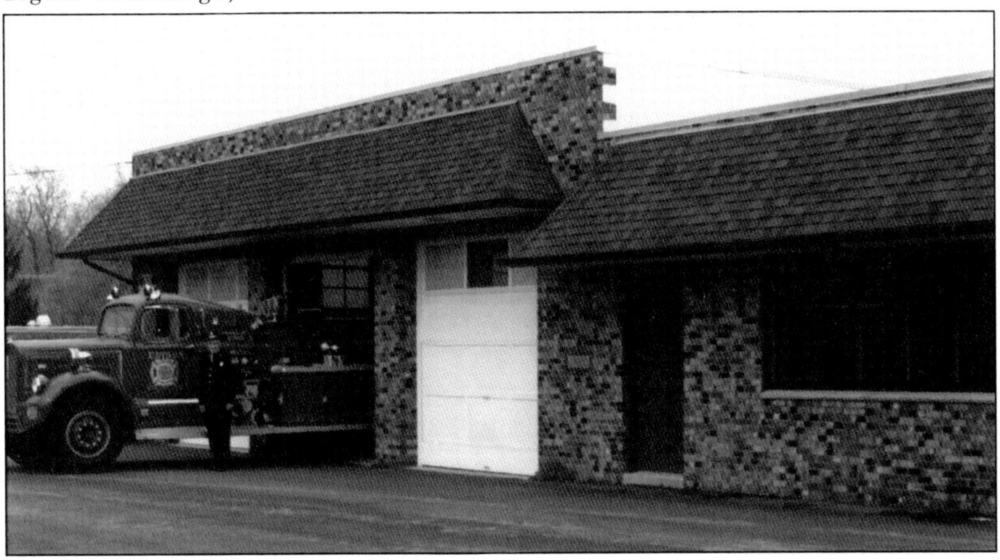

H. K. volunteer firefighter John Pencak is standing next to old engine No. 3, a 1956 White made by TASC, in front of the newly erected Edison Exempt Building. Members of the Edison Exempt Fireman's Association built this structure after receiving a donation of land directly behind the H. K. Volunteer Firehouse on New Dover Road from Edison Township. This brick structure, which was completed in November 1982, has a meeting hall with three bays to store vintage fire apparatus. The building was dedicated in April 1983. (Courtesy of John Pencak.)

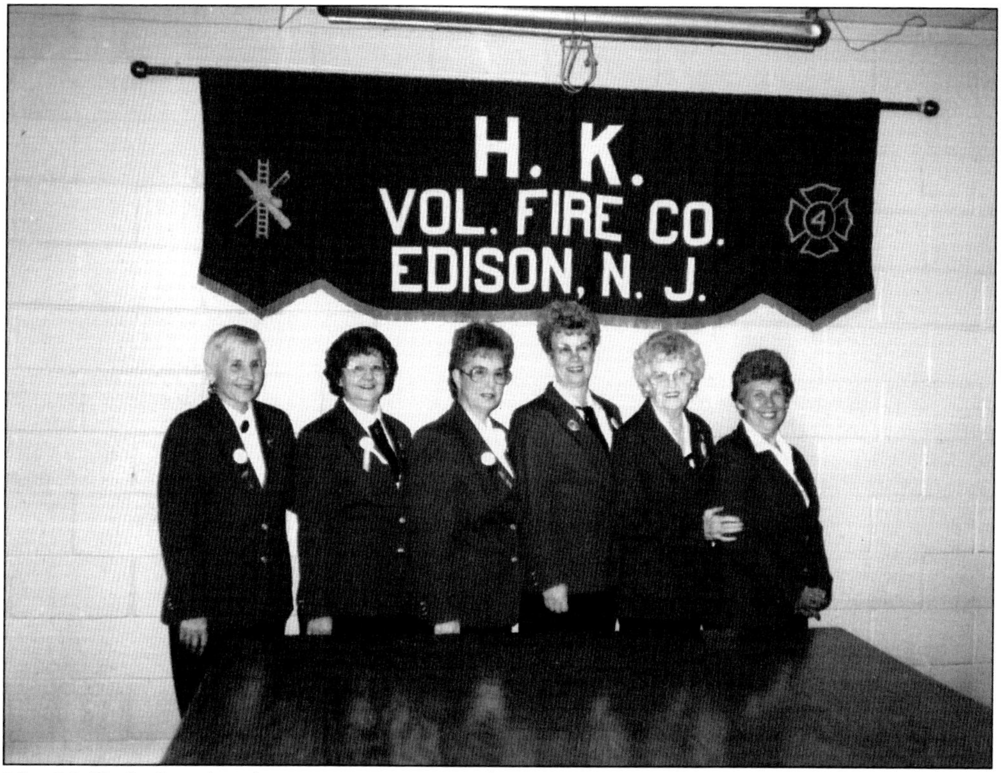

The H. K. Ladies Auxiliary was incorporated on April 27, 1942, with the primary purpose of supporting the firefighters and aiding the community during emergency situations. The membership, which outlasted the other ladies auxiliaries in Edison, remained active up until the late 1990s. Pictured above (from left to right) are Heidi Macyszyn, Ann Lichowid, Amy Zavis, Ada Enfield, Dorothy Fairbanks, and Collette Wolent. (Courtesy of Ada Enfield.)

H. K. VOLUNTEER FIRE COMPANY

EX - CHIEFS

Name	Years	Name	Years
Edwin Minew	1926	Michael D'Alasandro	1959-61
Charles Bott	1927-33	Davis Cherick	1962-63
Lloyd Krommes	1935	Charles Whitecavage	1964-66
Charles Bott	1936	John Pencak	1967-69
O. Goodrow	1937-39	Walter Karlo	1970-72
Charles Schmidt	1940-41	Ludwig Roell	1973-75
E. Ehrman	1942	Patrick Novia	1976-77
George Yaeger	1942-45	Steven Mozsgae	1978-79
John Lamb	1946	Robert Hallbauer	1979-81
Evans Lindquist	1947	Edward Koehler Sr	1982-83
William Pryor	1948	Gerald Sorrentino	1984-85
Evans Lindquist	1949	Richard Buda	1986-87
William Fairbanks	1950	Robert Brunt	1988-89
Frank Stanziola	1951-53	Edward Koehler	1990-01
Evans Lindquist	1954	Robert Brunt	2001-06
Alphonse Novia	1955	John Pencak	2007
George Hann	1956-58		

Listed above are the chiefs that have led H. K. Volunteer Fire Company throughout the years.

Five

Oak Tree Volunteer Fire Company

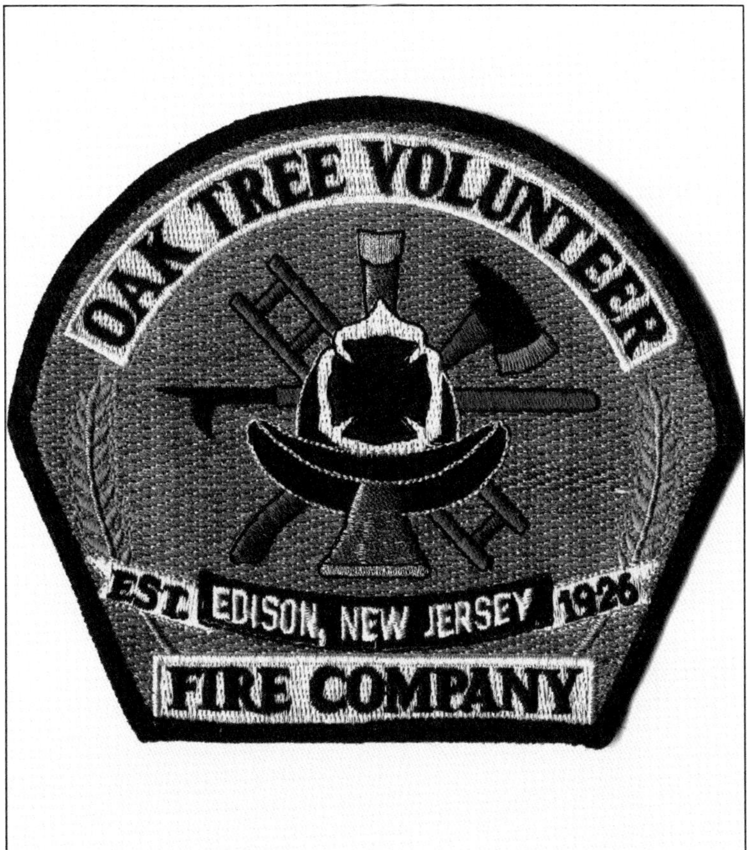

Sometime between 1925 and 1926, a group of 20 men gathered together and formed the Oak Tree Terrace Volunteer Fire Company. The company provided fire protection for the Oak Tree portion of Raritan Township by using water buckets and grass-beaters. Grass-beaters (made of a thick piece of leather attached to the end of a pipe) were used to "beat out" field fires. The company received alarms by the sound of someone banging an object on the nearby railroad tracks.

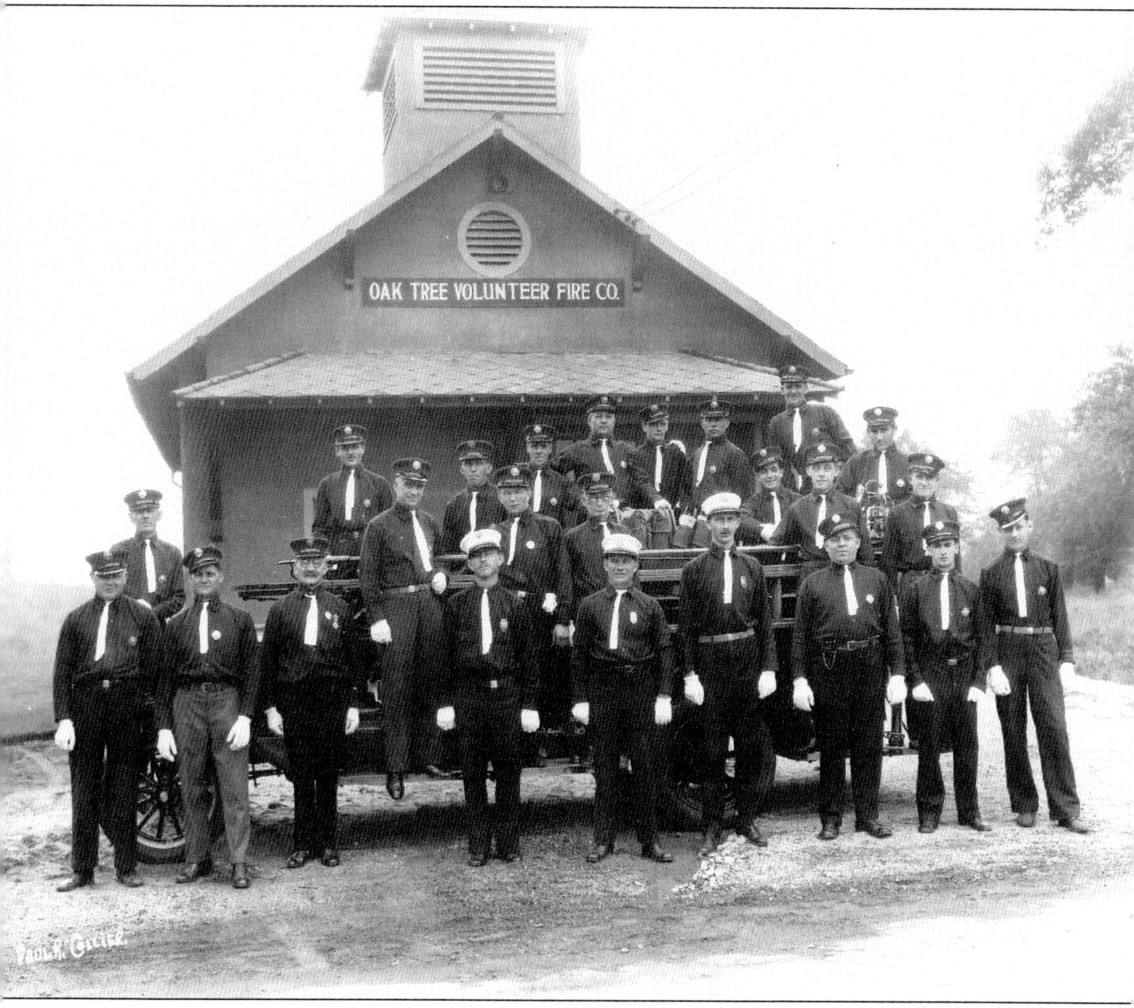

In February 1927, Oak Tree residents elected the first board of commissioners for Fire District No. 5. Once the commissioners were elected, they immediately purchased a Ford Model T fire engine and 500 feet of hose, which were delivered four months later. Up until the construction of the firehouse, the fire engine was kept in the Kirkpatrick's family barn. In June 1927, the commissioners purchased land on Oak Tree Road near the Lehigh Valley Railroad crossing for $350. The construction of a new firehouse was completed within one year and was dedicated on May 30, 1928. This early photograph shows the original members standing in front of their new firehouse in the late 1920s. Pictured from left to right are (first row) William Jerema, Peter Tesauro, Henry Wassenberg, Percy Vroom, Chief Leonard Sanford, unidentified, unidentified, Samuel Kirkpatrick, and unidentified; (second row) Orville Freeman, Robert Krog, Charles Hamaker, James Kirkpatrick Sr., Masimo Quagliariello, John Deering Jr., James Giles, and Louis Kraus; (third row) William Reed, Abram Hamilton, unidentified, unidentified, Alexander Kirkpatrick, unidentified, and Charles Freeman Jr. (Courtesy of Oak Tree Volunteer Fire Company.)

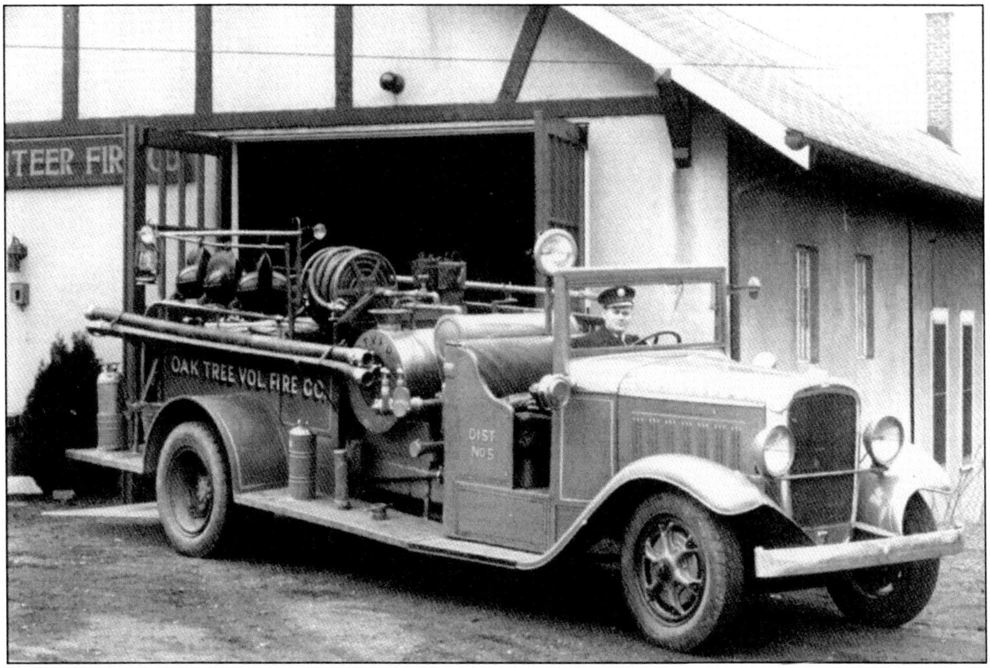

In 1932, the board of fire commissioners for Fire District No. 5 purchased a new Studebaker fire engine to replace the old Ford Model T. Notice the soda-acid extinguisher on the rear bumper. This type of extinguisher was invented in the 19th century, which contained a water-sodium bicarbonate solution with a suspended vial of sulfuric acid. To activate the extinguisher, the user would turn it upside down so that the internal vial would mix with the solution to produce carbon dioxide gas, thereby pressurizing water out through the nozzle.

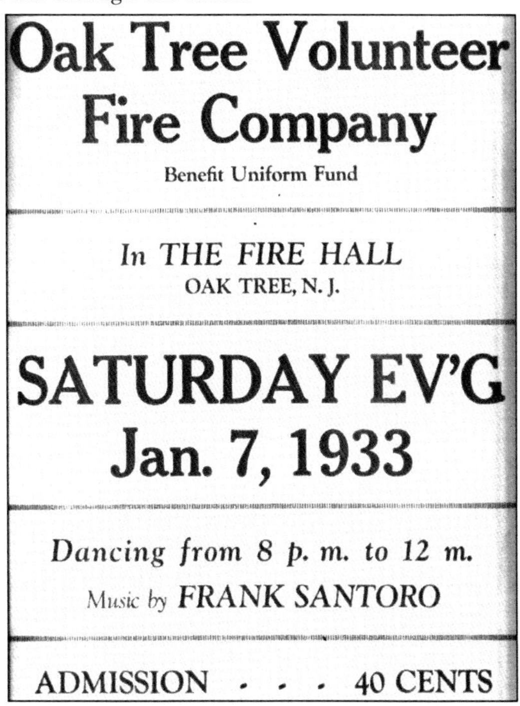

The Oak Tree Volunteer Fire Company hosted dances in the firehouse in order to raise money. For this particular fund-raiser, profits helped to pay for the company's dress uniforms.

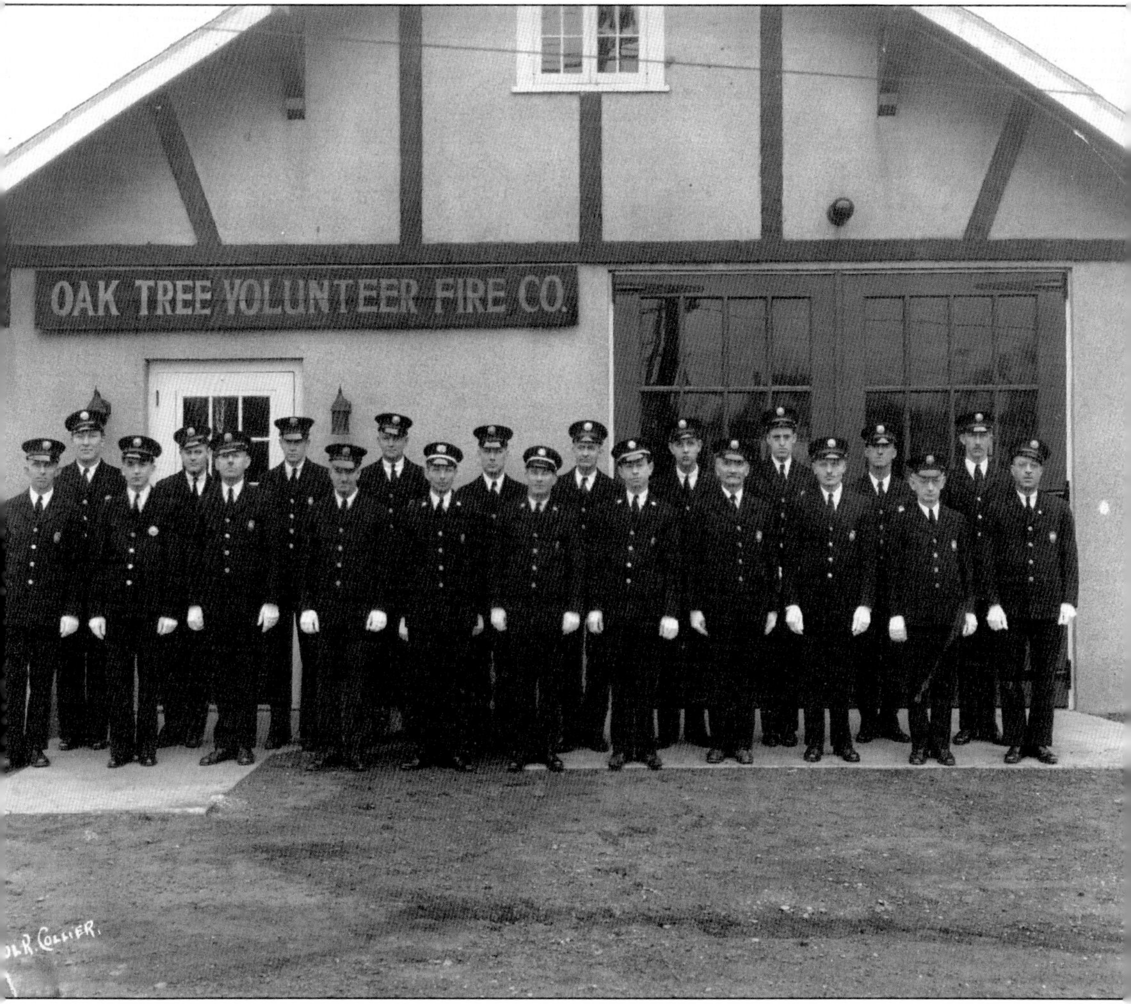

This photograph of the Oak Tree Volunteer Fire Company membership was taken in 1939, just after expansion and renovations to the firehouse. For many years, the Oak Tree Library was located within this firehouse. Mrs. Wier, the local librarian, organized books within a metal filing closet in the corner of the firehouse's dance hall. Pictured from left to right are (first row) Abraham Hamilton, Philip Quagliarello, William Reed, John Deering Sr., Louis Kraus, Chief Leonard Sanford, Percy Vroom, Henry Wassenberg, William Jerema, Samuel Kirkpatrick, and Patrick Triola; (second row) George Ulrich, Robert Bidmead, Charles Freeman Jr., James Giles, Harold DeLisee Sr., Orville Freeman, unidentified, Norman Freeman, unidentified, and Russell Hamilton. (Courtesy of Walter Ulrich.)

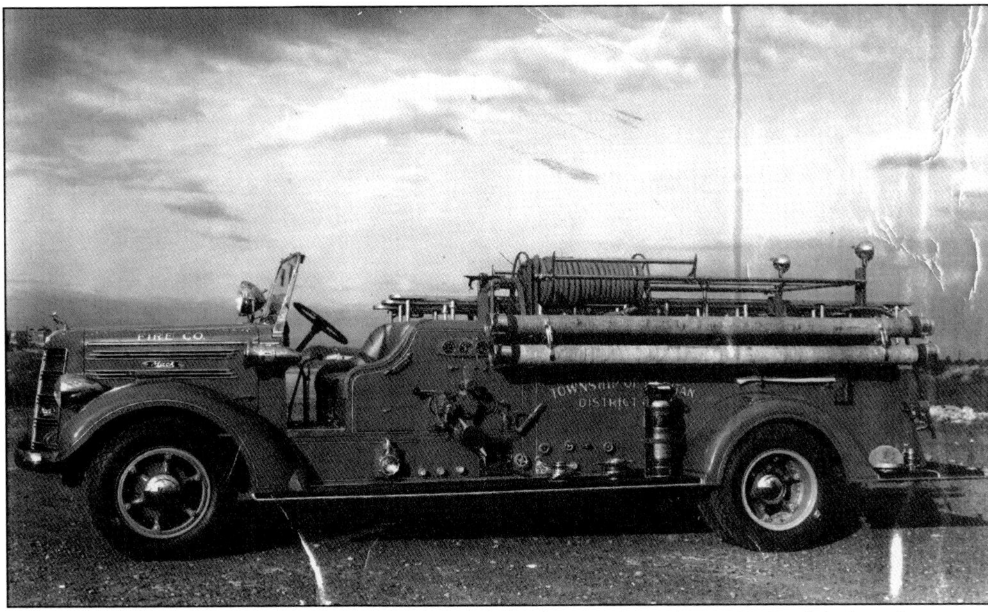

A brand new 1941 Mack fire engine was purchased to replace the 1932 Studebaker that met a sad fate earlier that year. While responding to a barn fire, the Studebaker collided with another vehicle. Although the engine eventually arrived at the fire, damages were beyond repair. (Courtesy of Oak Tree Volunteer Fire Company.)

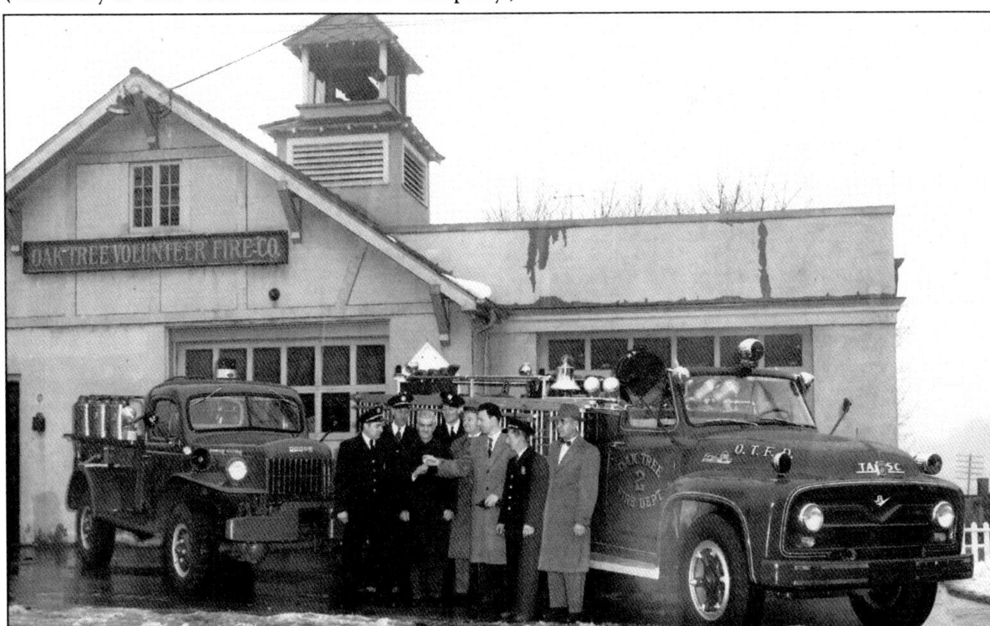

On February 2, 1955, Oak Tree Volunteer Fire Company received a new 1954 Dodge Power Wagon brush truck and a new 1954 Ford fire engine. This important photograph shows the fire commissioners of Fire District No. 5 officially turning over the apparatus keys to the firefighters. Pictured from left to right are Chief Charles Wojciechouski, Assistant Chief Irving Day, John Deering Sr., Elmer Magee, John Needham, Nicholas Alicino, firefighter Rudolph Bjorkland, and Arthur Cadavero. (Courtesy of Walter Ulrich.)

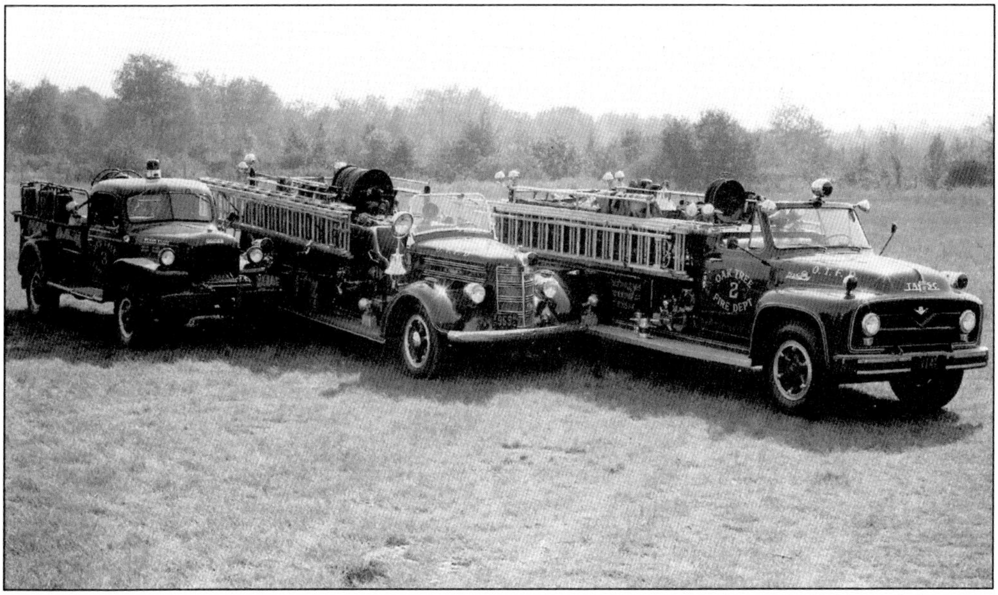

This 1955 photograph shows the 1941 Mack in between the new brush truck (left) and the new Ford fire engine with a typical 1950s Oak Tree setting in the background. The brush truck, a 1954 Dodge Power Wagon that cost $4,000, was equipped with a small pump with a reel of hose mounted in the rear for extinguishing brush fires. The fire engine, a 1954 open-cab Ford that cost almost $14,000, was built by TASC. (Courtesy of William Prairie.)

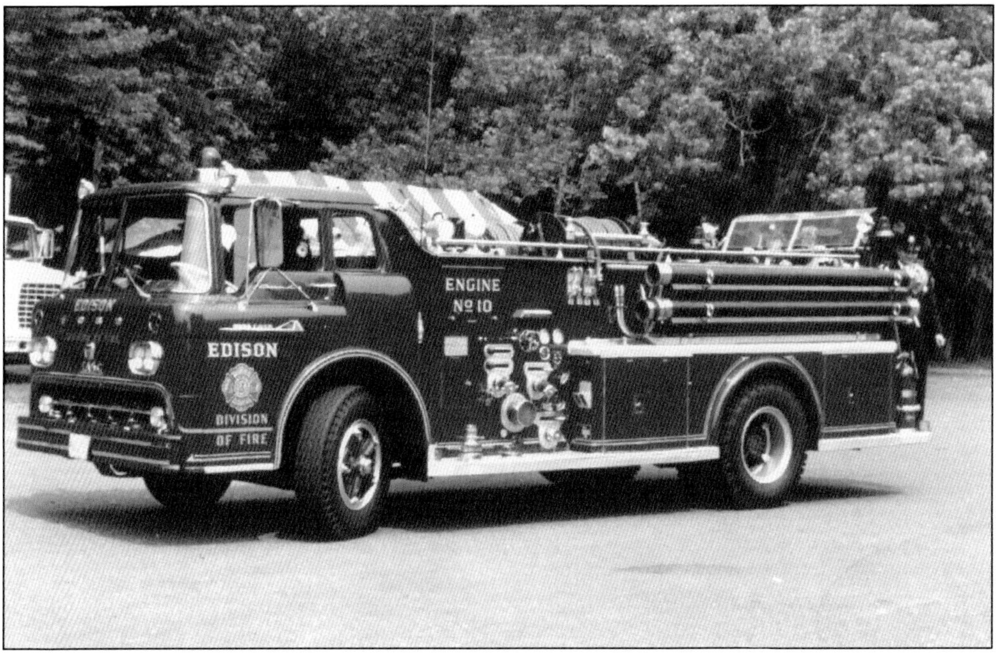

After the elimination of the fire districts and their associated fire commissioners, the newly formed Division of Fire purchased this 1959 Ford fire engine from TASC for the Oak Tree Volunteer Fire Company. This enclosed-cab Ford was equipped with a 500-gallon booster tank and a Hale pump capable of discharging 750 gallons of water per minute. (Courtesy of Edison Volunteer Fire Company No. 1.)

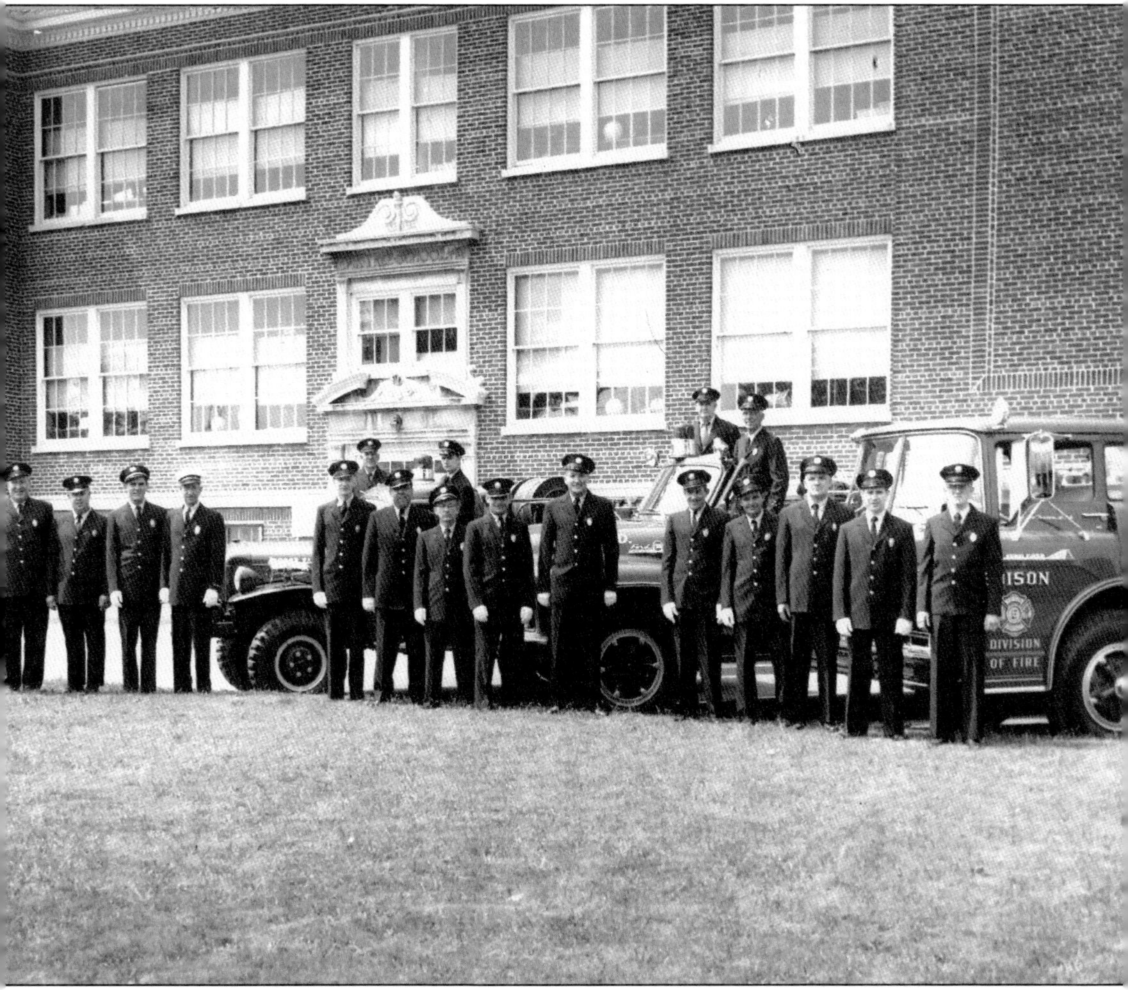

The new 1959 Ford fire engine replaced the old 1941 Mack fire engine, which was involved in an accident a year or two prior to this photograph. While responding to a structure fire involving an old two-story house in Potters, the Mack rolled over onto its side. The driver, who was the only firefighter in the vehicle, sustained only minor injuries. This 1960 photograph shows the members of Oak Tree Volunteer Fire Company posing with the 1954 Dodge Power Wagon, the 1954 Ford TASC, and the new 1959 enclosed-cab Ford TASC fire engine in front of the Oak Tree School. Pictured from left to right are firefighters George Ulrich, John Peters, John Wasko, Irving Day, George Michaelson, Franklyn Bardecker Sr., Henry Tempe Sr., Rudolf Bjorkland standing in front of Albert Havrilla, John Jerema, Elmer Magee, Hampton Yates in front of Charles Freeman, Victor Almes in front of Abram Hamilton, Harry Parsons, Bernard Mulhearn, and James Hamilton. (Courtesy of Oak Tree Volunteer Fire Company.)

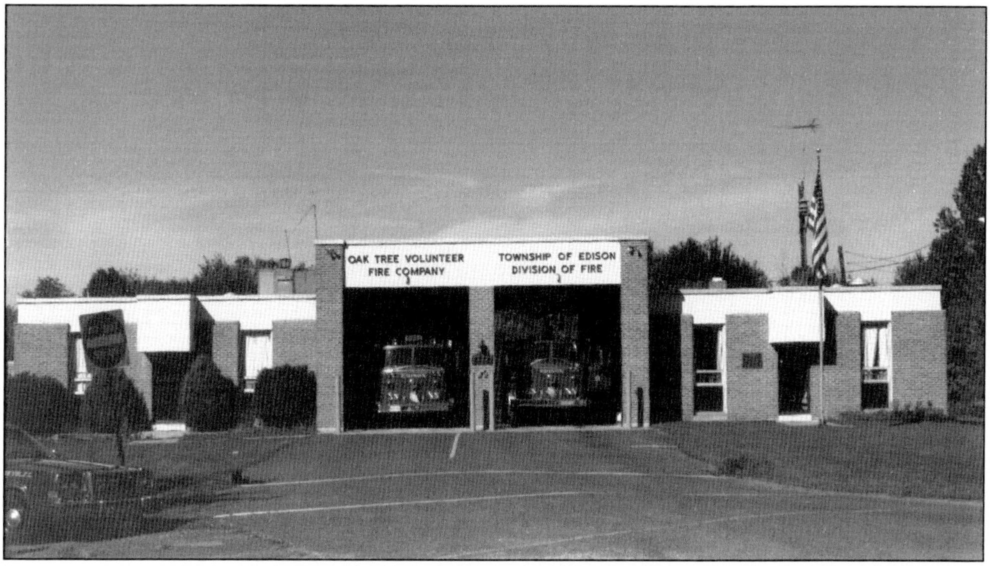

During the late 1960s, Oak Tree chief Henry Tempe Sr. convinced Joseph Simon, supervisor of the Edison Division of Fire that a new firehouse should be built. A township lot about 500 feet east of the original firehouse was secured for the new location of the firehouse. In October 1970, construction began, and by August 1971, the new firehouse on Beverly Street was completed. This is the only firehouse in Edison that was built with separate quarters to accommodate both career and volunteer firefighters. (Courtesy of Scott McCloughan.)

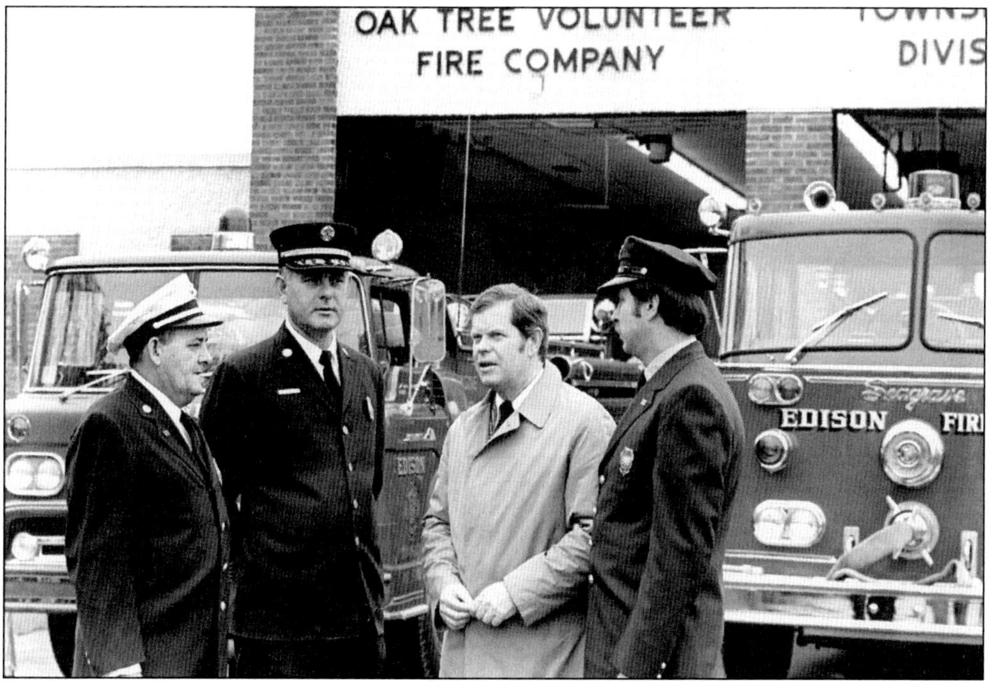

On October 29, 1972, the new firehouse on Beverly Street was dedicated. Pictured from left to right are Oak Tree volunteer chief Victor Almes, Division of Fire chief H. Ray Vliet, Mayor Bernard Dwyer, and Oak Tree president Michael Hester. (Courtesy of Oak Tree Volunteer Fire Company.)

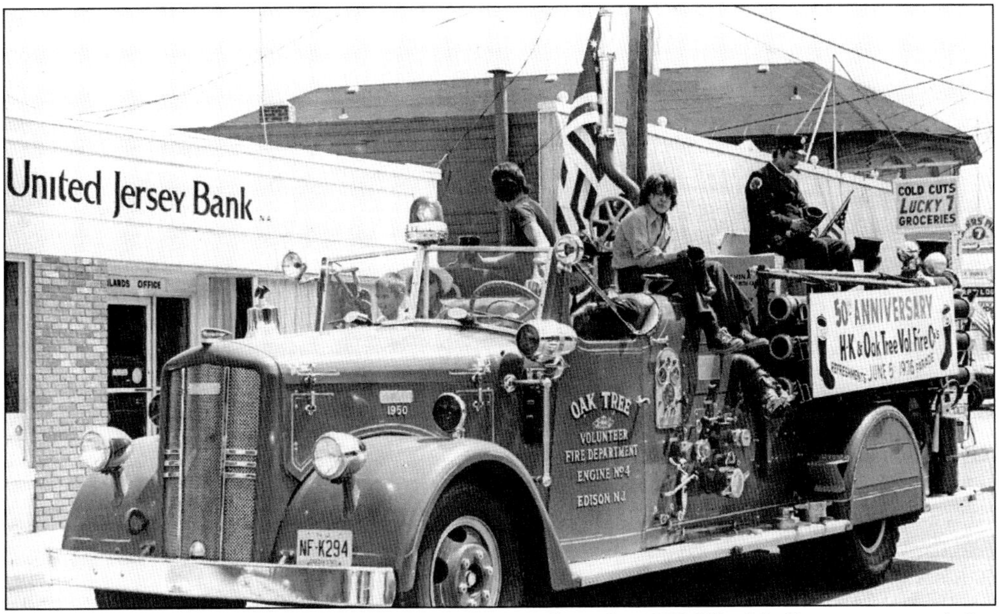

In 1975, the Oak Tree Volunteer Fire Company purchased a 1950 Ward LaFrance open-cab fire engine from South Plainfield to use for parades only. This photograph shows firefighter Franklyn Bardecker Sr. driving the Ward LaFrance in South Plainfield's 1976 Memorial Day parade with his wife, Carol, sitting in the passenger seat holding their son Timothy. Young Terry and Steven Bardecker are riding in the rear with firefighter Timothy Finne as he casually smokes his cigar. (Courtesy of Oak Tree Volunteer Fire Company.)

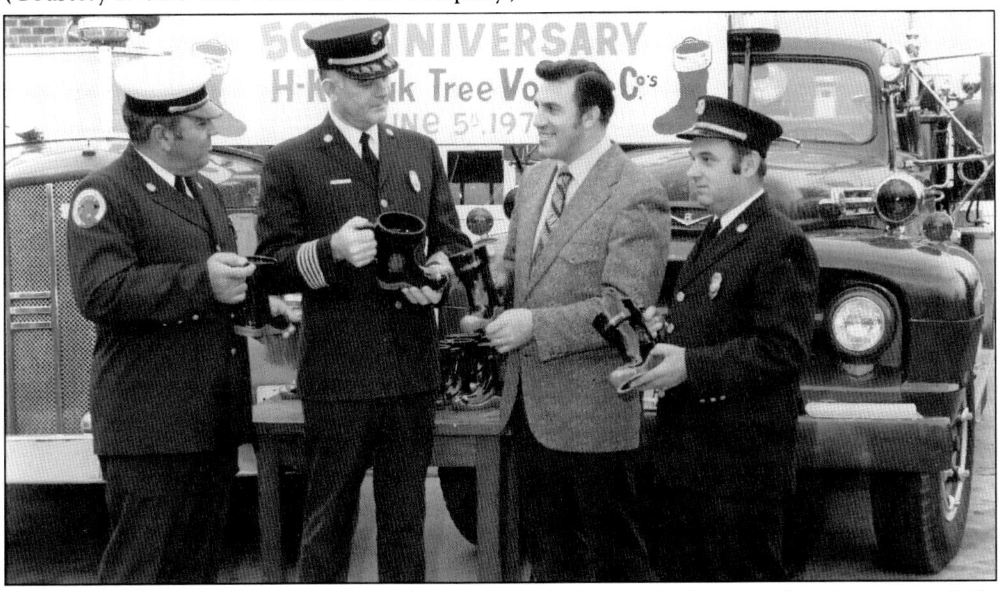

This 1976 photograph was taken in front of the Oak Tree Volunteer Fire Company's 1950 Ward LaFrance and the H. K. Volunteer Fire Company's 1952 Ford fire engine during their 50th anniversary celebration. From left to right, Chief Henry Tempe Sr. (Oak Tree Volunteer Fire Company), Division of Fire chief H. Ray Vliet, Mayor Thomas Paterniti, and Chief Ludwig Roell (H. K. Volunteer Fire Company) are seen discussing the success of their unique fire souvenir boot mug. (Courtesy of H. Ray Vliet.)

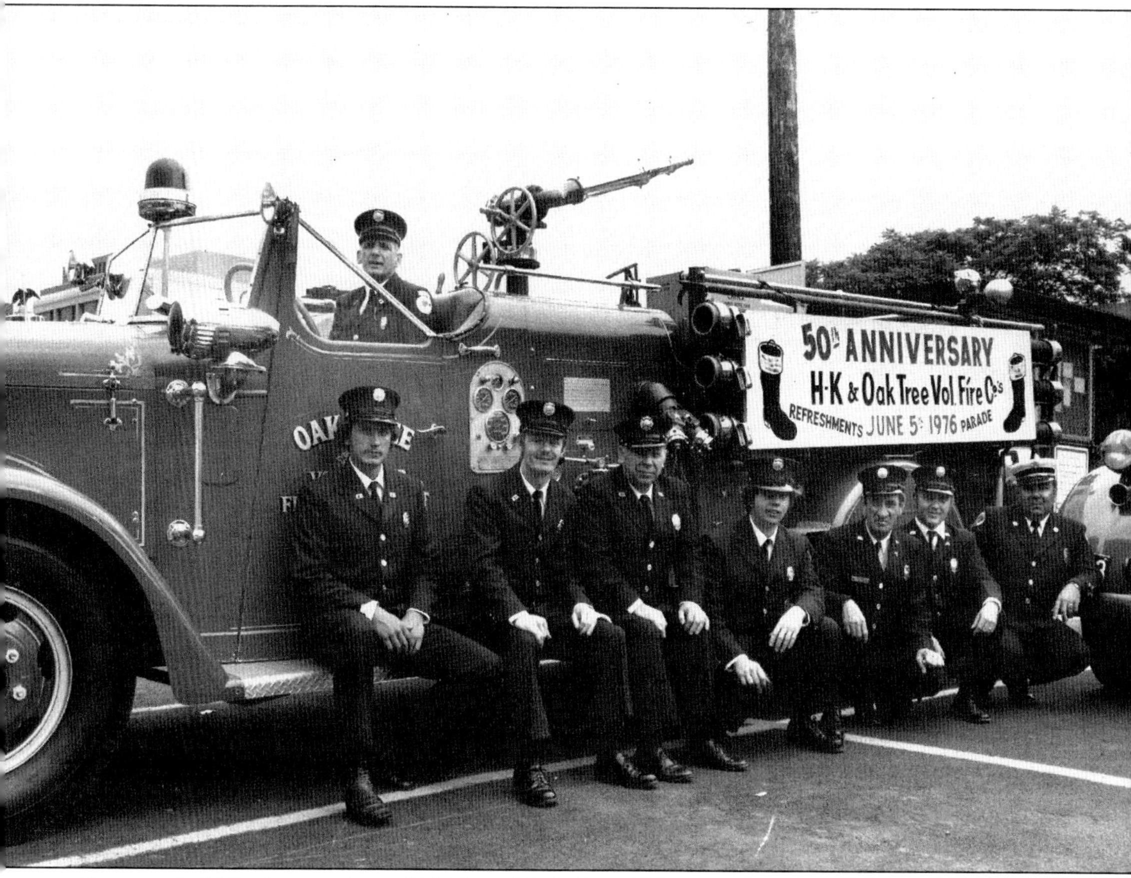

On June 5, 1976, the members of H. K. and Oak Tree Volunteer Fire Companies united together to celebrate a joint 50th anniversary celebration. A big parade, which featured actress and former Edison resident Susan Sarandon as the parade queen as well as actress and former Edison resident Gail Fisher as the parade grand marshal, ended at the James Madison Primary School parking lot on New Dover Road. During the festivities, some of the members from both companies posed in front of Oak Tree's Ward LaFrance. From left to right are firefighters Joseph Finan, Gary Shelhimer, Raymond Volker Sr., Dennis Phelan, Edward Koehler Sr., Eugene Enfield Sr., Henry Tempe Sr., and Franklyn Bardecker Sr. sitting behind the wheel. (Courtesy of Eugene Enfield Sr.)

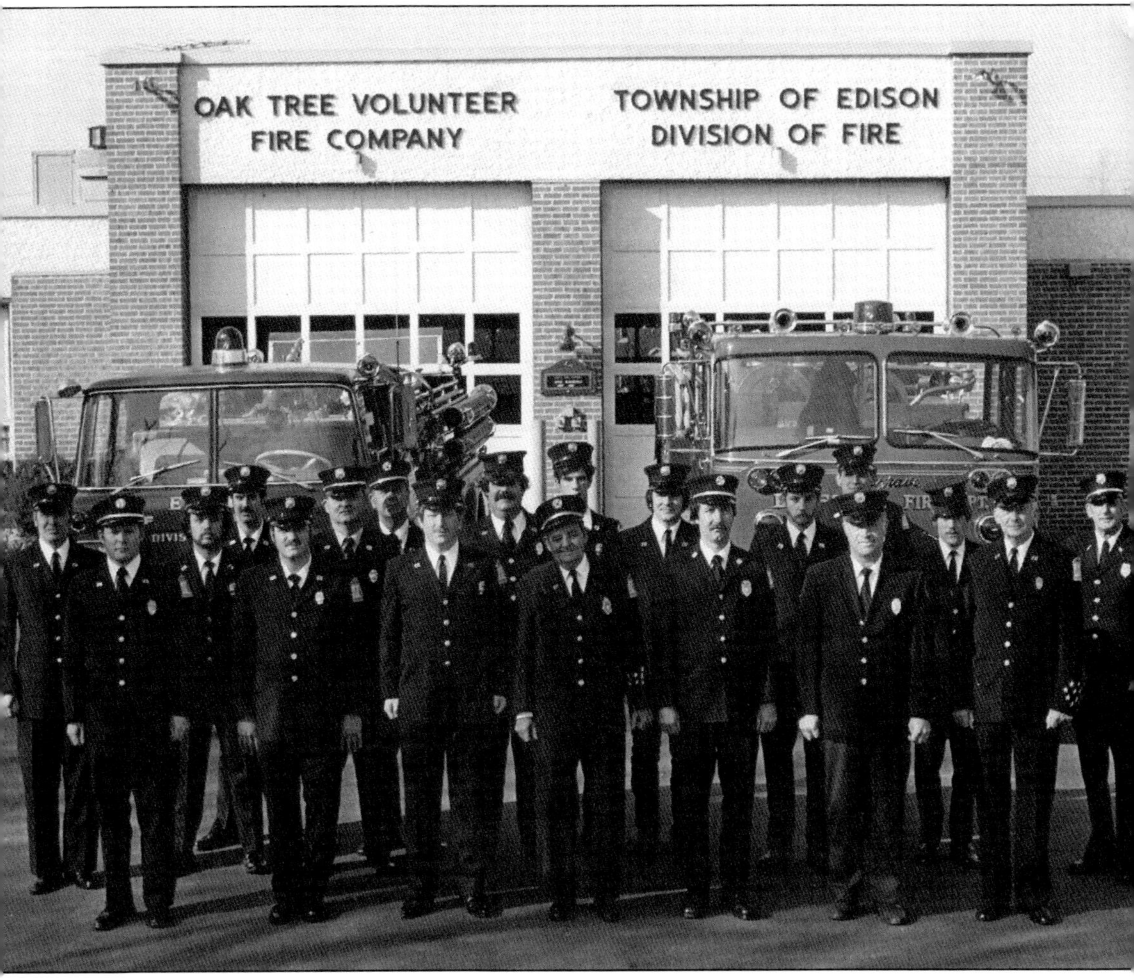

The members of the Oak Tree Volunteer Fire Company are gathered together for a formal photograph in front of the firehouse on Beverly Street in 1979. Pictured from left to right are (first row) Thomas Weeks, James Egan, Rodney Masterson, Victor Almes, Chief William Prairie, John Smith, George Michaelson, and Franklyn Bardecker Sr.; (second row) Vincent Kraus, Jeff Thomas, Jules Kwitkokski, Joseph Czech, Steven Bardecker, John Hansen, and Douglas Walp; (third row) Scott Lewis, John Peters, George Ulrich, and Terry Bardecker. (Courtesy of Oak Tree Volunteer Fire Company.)

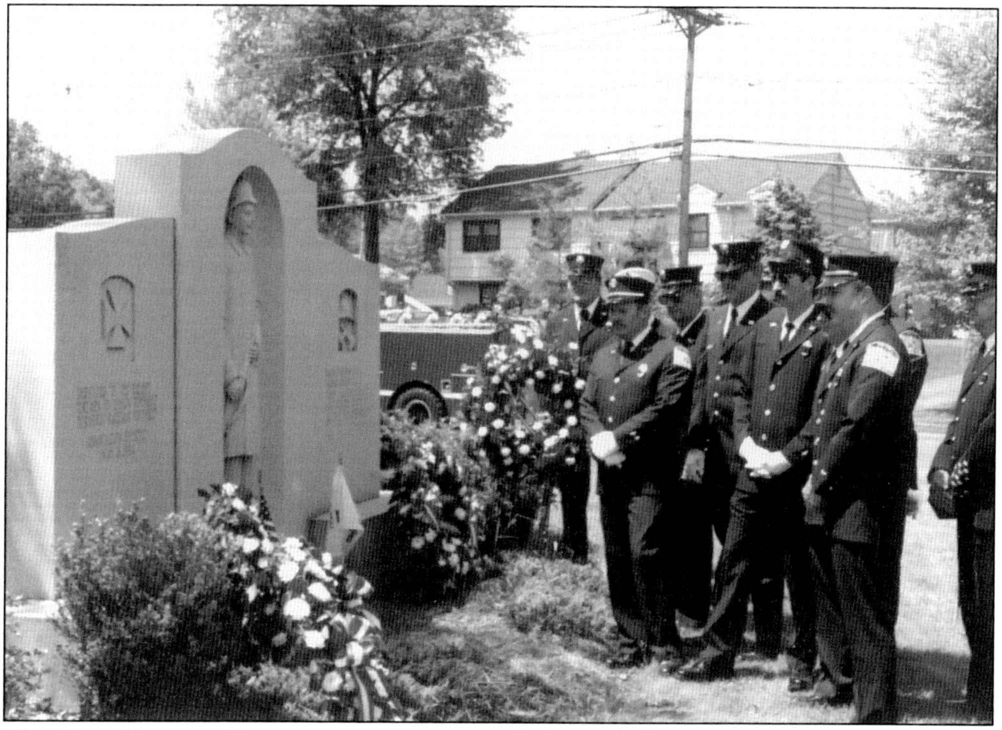

The Edison Fireman's Exempt monument, originally located off Plainfield Avenue near Simpson Avenue, was moved to Grove Avenue near Library Place when the old town hall was sold in 1980. Shown above are members of Oak Tree Volunteer Fire Company honoring the exempt members of Edison Township on Memorial Day 1986. (Courtesy of Oak Tree Volunteer Fire Company.)

OAK TREE VOLUNTEER FIRE COMPANY

EX - CHIEFS

Edward Deering	1927	Henry Tempe Sr.	1974-76
Leonard Sanford	1928-46	John Smith	1977-78
Charles Wojciechouski	1947-55	William Prairie	1979-82
Irving Day	1955-64	Thomas Weeks	1982-84
Harry Parsons	65-66	William Prairie	1985-89
Henry Tempe Sr	1967-70	James Mahoney	1990-99
Victor Almes	1971-72	William Prairie	2000-01
John Peters	1973	Paul Jean	2002-03
		Kenneth DiFrenza	2004-07

Listed above are the chiefs that led Oak Tree Volunteer Fire Company throughout the years.

Six

EDISON DIVISION OF FIRE

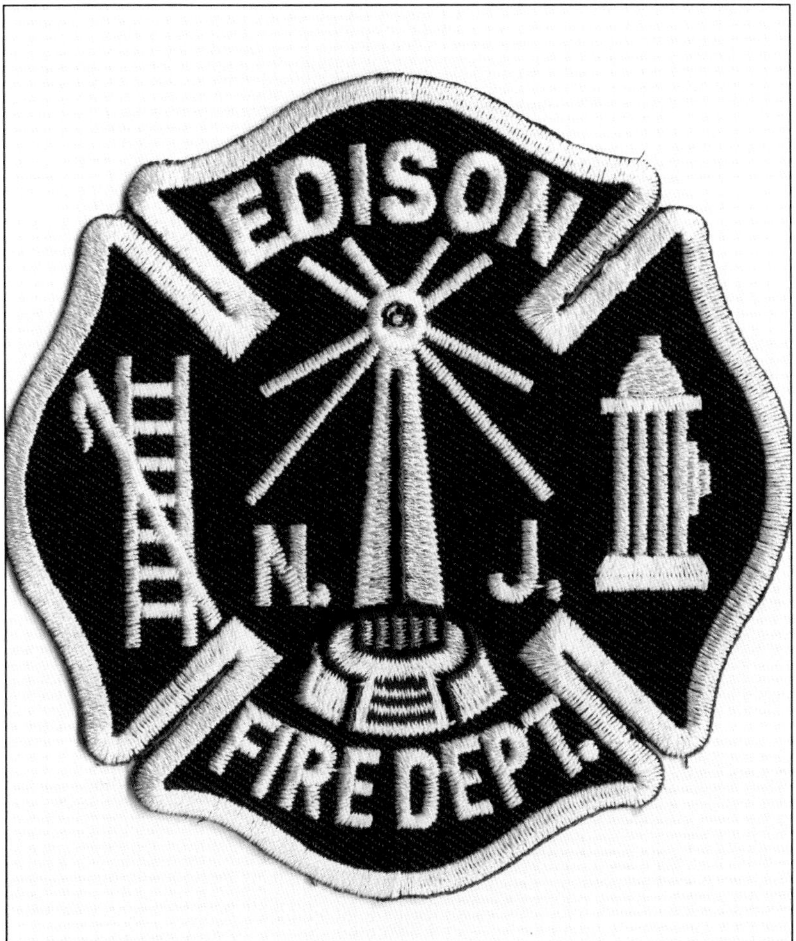

On January 1, 1958, Edison Township changed its form of government, abolished fire districts, and created the Edison Division of Fire. Newly elected mayor Anthony Yelencsics appointed Joseph Simon as the division's supervisor with administrative authority over all five volunteer companies and all career firefighters from District No. 1 and District No. 3. The patch at right is only worn by those who serve as career firefighters for the Edison Division of Fire.

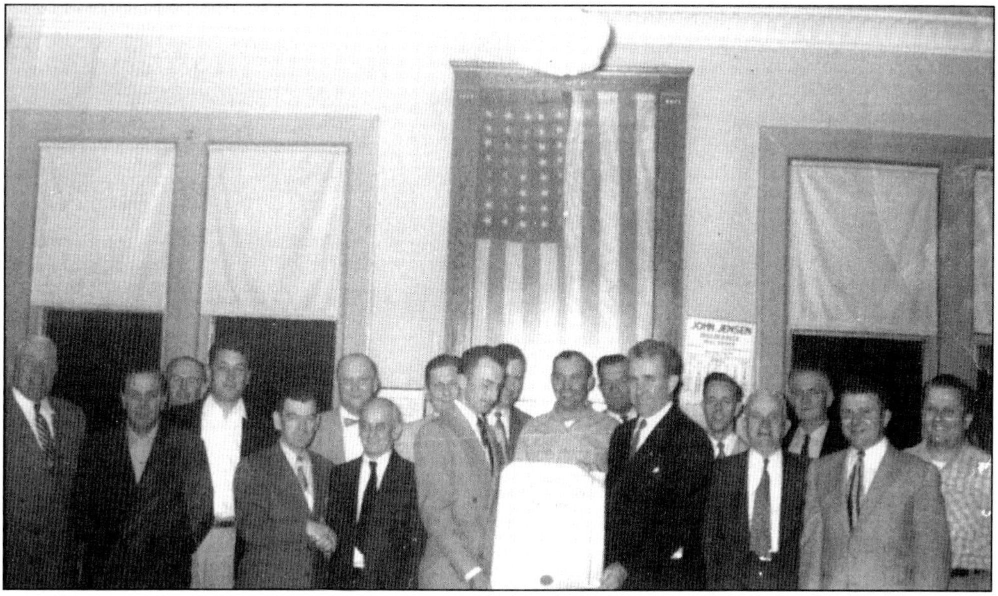

In 1955, career firefighters from Fire Districts No. 1 and No. 3 united to form a labor union. These men contacted the International Association of Fire Fighters (IAFF), which was well known for its global strength and affiliation with the American Foundation of Labor and Congress of Industrial Organization (AFL-CIO). After following proper procedures, the firefighters received their charter on April 1, 1955, and became officially known as the Edison Township Fire Fighters Association, Local No. 1197. This photograph shows union charter members receiving their charter during a special ceremony on the second floor of Raritan Engine Company No. 2's firehouse in Clara Barton. (Courtesy of William Schnieder, photograph by Calvin Latham.)

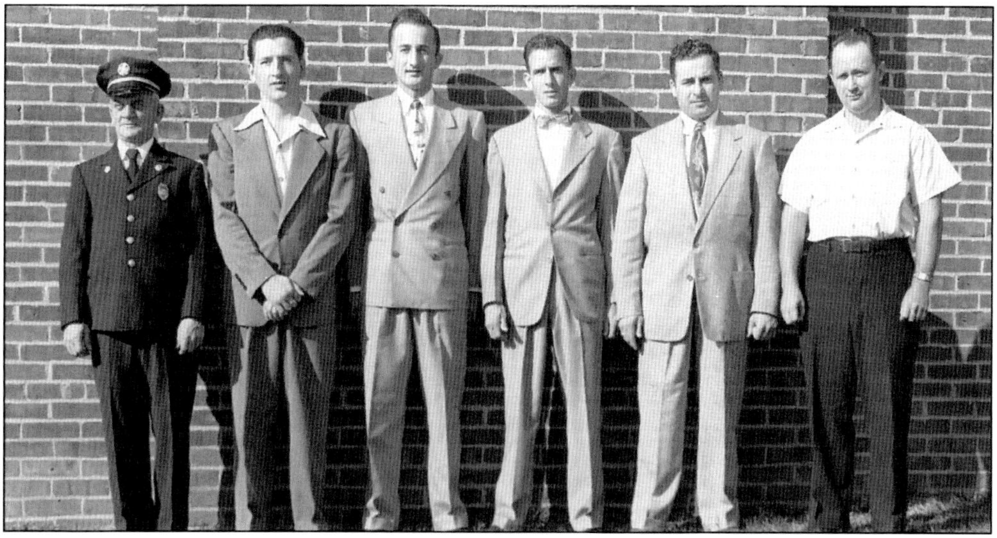

In 1955, five township firefighters, one from each fire district, attended a four day advanced firefighting course at Fort Dix with expenses paid by the Raritan Township Civil Defense Council. Chief Engineer Arthur Latham (left), chairman of the civil defense fire protection committee, is standing next to (from left to right) Steven Burylo (Midwood), Donald Dudics (Clara Barton), Calvin Latham (Piscatawaytown), Charles Wojciechowski (Oak Tree), and Martin Dige (Menlo Park). (Courtesy of Walter Ulrich.)

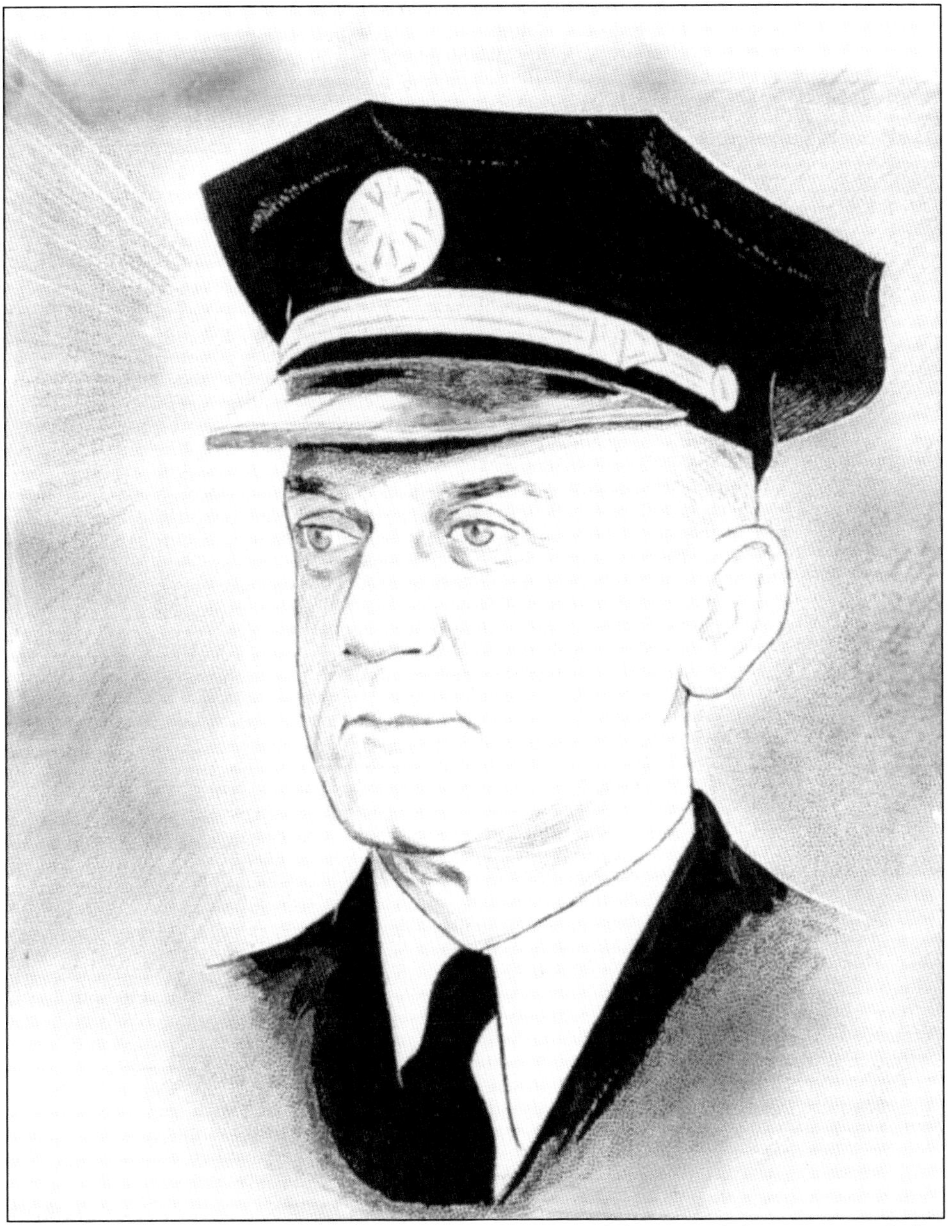

The fire commissioners of Raritan Fire District No. 1 agreed that a paid driver was needed and, in 1926, Arthur W. Latham was hired and became the first career firefighter of Edison (known as Raritan Township at that time). Within a short period of time, Latham earned the official title of chief engineer of Fire District No. 1 and had an assistant, William Fisher, to help him maintain the firehouse, equipment, and most importantly, extinguish fires. By the mid-1950s, 15 firefighters were working in District No. 1 under Latham's command. On January 1, 1958, Edison Township abolished the fire districts and created the Edison Division of Fire. Shortly after the change, Latham began planning for his retirement for June of that year. Unfortunately, Latham passed away on May 4, 1958—just one month prior to his retirement. Chief Arthur Latham will always be remembered and honored for his service and dedication, and especially for providing the foundation of Edison's career fire service. (Courtesy of Edison Division of Fire.)

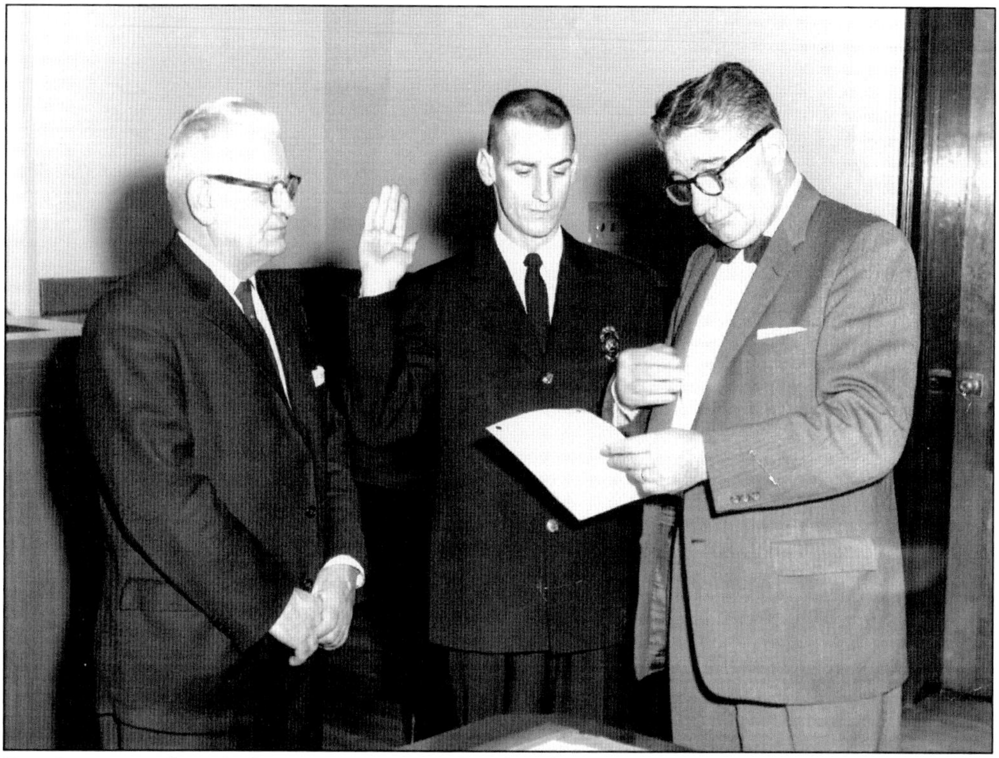

Fire Supervisor Joseph Simon witnesses Richard Latham being sworn in as a career firefighter in February 1962. Latham raises his right hand while township clerk Arthur Tucker dictates the official oath of office. Latham's annual salary was $4,200. (Courtesy of Brian Latham.)

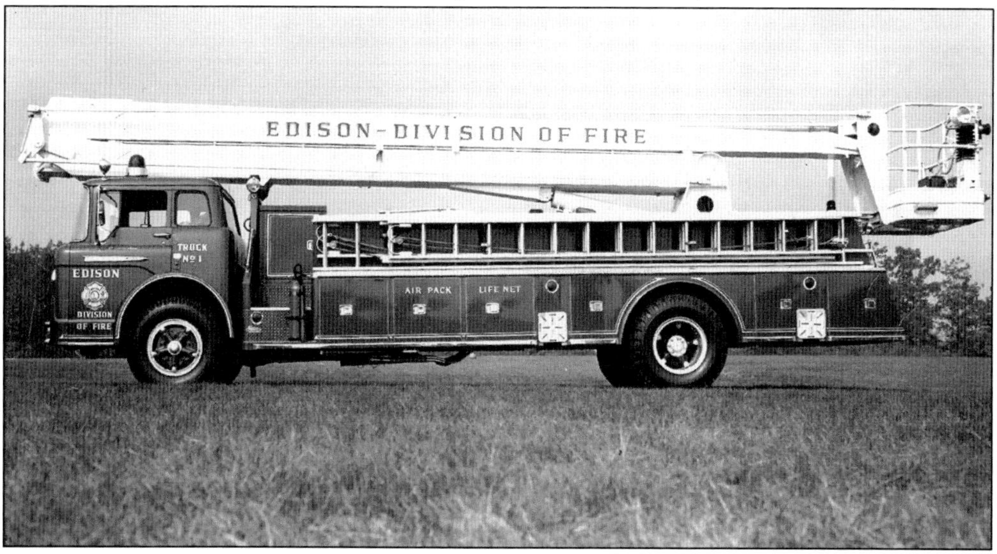

The Edison Division of Fire was the first fire department east of the Mississippi River to purchase a fire truck equipped with an aerial platform. This 1961 Ford Pitman Snorkel (designated as truck 1) was one of a kind; it had a 65-foot articulating boom that served the needs of Edison Township as well as those of neighboring townships. Therefore, truck 1 was used quite often throughout the years. (Courtesy of Brian Latham.)

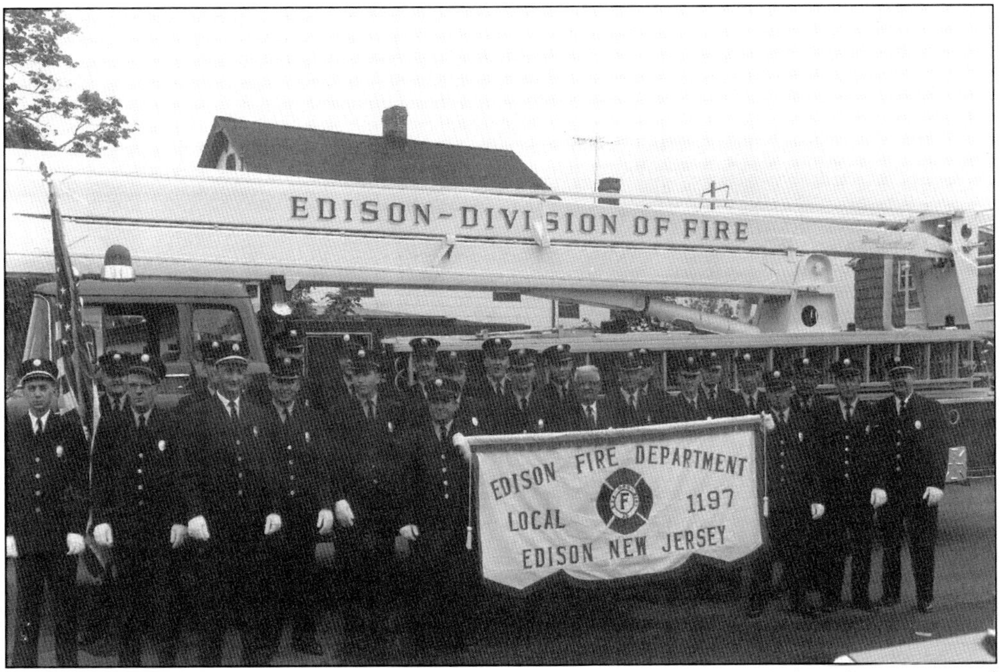

The members of the Edison Division of Fire pose in front of their new Snorkel truck on Memorial Day 1962. (Courtesy of H. Ray Vliet.)

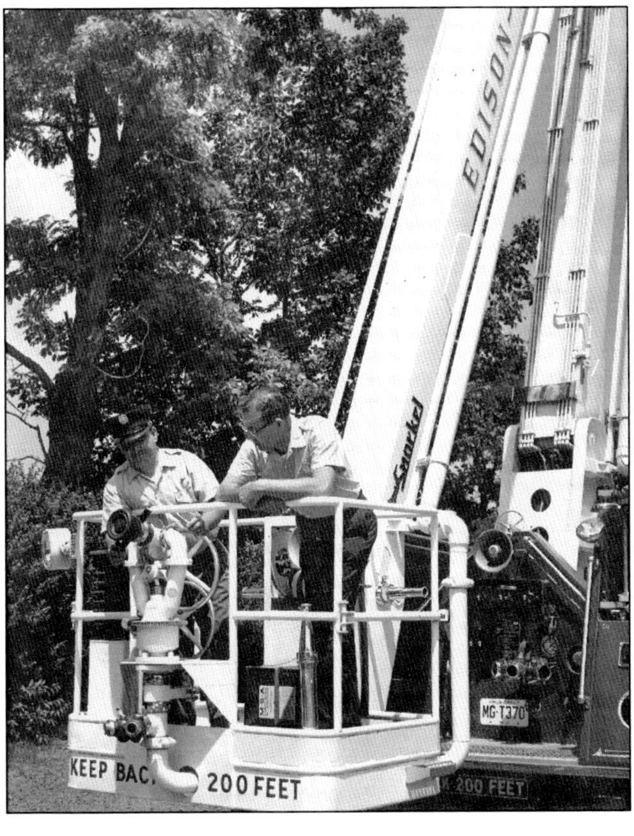

Firefighters James Colletto (right) and James Burke are testing the operation of the new Snorkel truck on July 10, 1962. (Courtesy of Edison Division of Fire.)

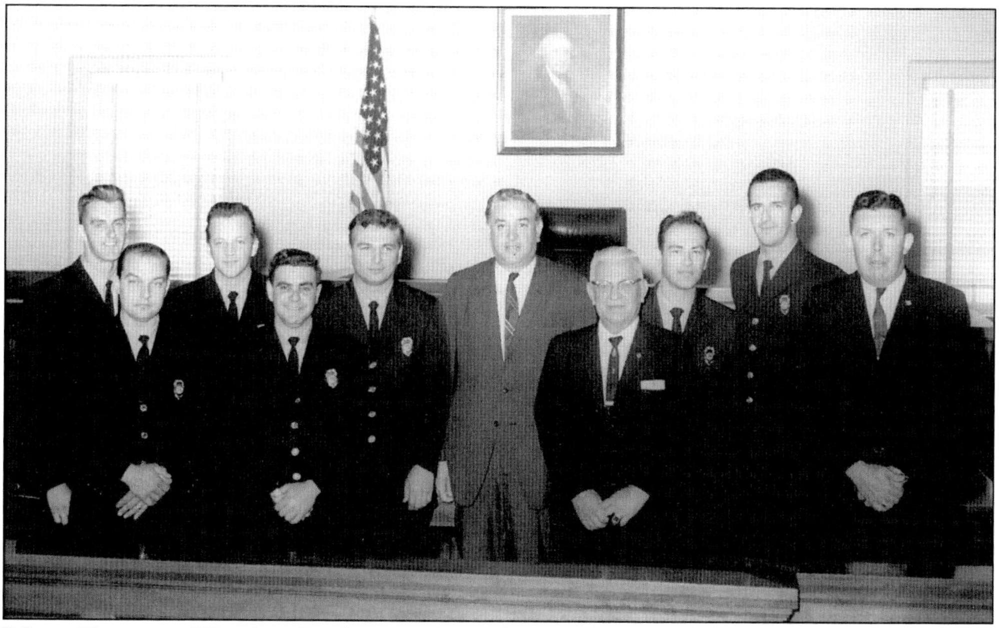

Edison Division of Fire hired eight new career firefighters on June 17, 1965. Pictured from left to right are Richard Borwegen, Joseph Stump, Robert Kearstan, Anthony Colletto, Alexander Danish, Mayor Anthony Yelencsics, Fire Supervisor Joseph Simon, John Dublanyk, Richard Rogan, and James Metz.

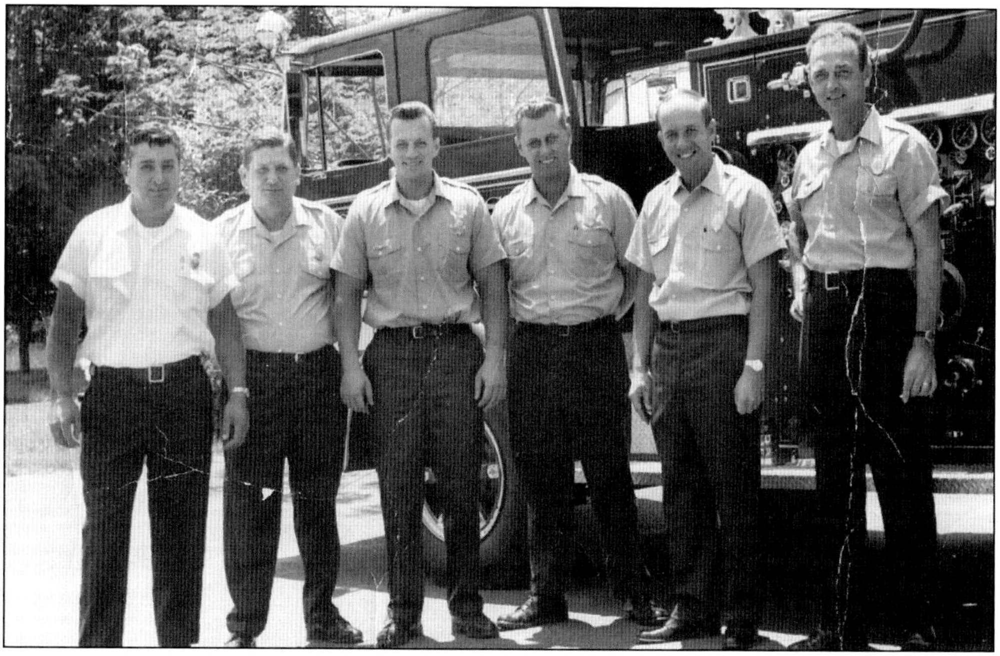

This photograph shows the members of D-Shift of the Edison Division of Fire gathered together in front of the new 1966 FWD fire engine at Fire Headquarters (now known as Station No. 3) on Amboy Avenue. Pictured from left to right are Capt. Milton Asprocolas, firefighters Joseph Sovart, Robert Kearstan Sr., David Jensen, Louis Sands, and Thomas Dalton. (Courtesy of Robert Kearstan Sr.)

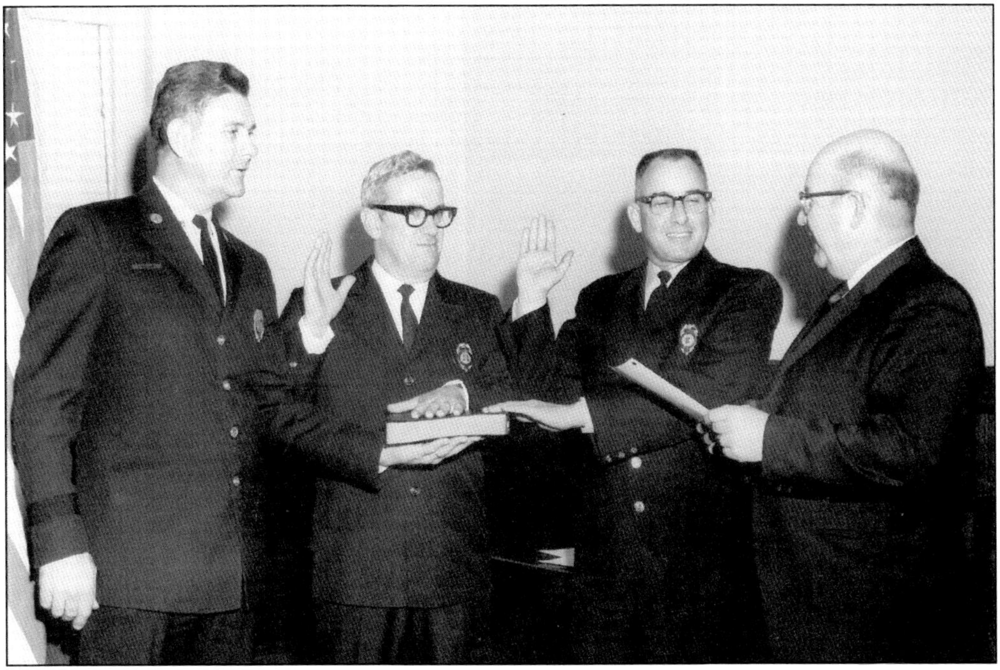

This 1970 photograph shows firefighter Edward Costello (second from left) being promoted to inspector and firefighter Calvin Latham being promoted to captain. Chief H. Ray Vliet is holding the Bible while township clerk Arthur Tucker dictates the oath.

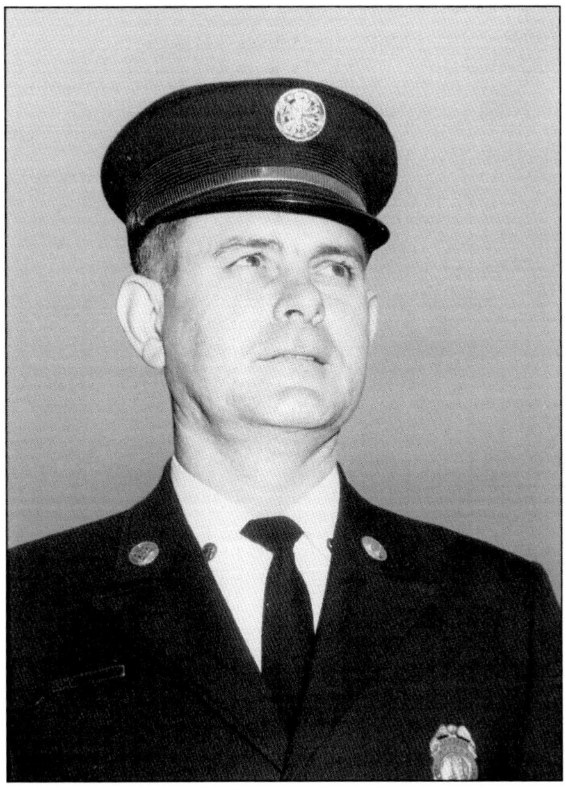

In 1970, Capt. H. Ray Vliet was offered the position of supervisor of fire to succeed Joseph Simon. He was interested, but the police and fireman's pension system did not recognize the "supervisor" title. The township modified the ordinance and he, therefore, became the first chief of the Edison Division of Fire.

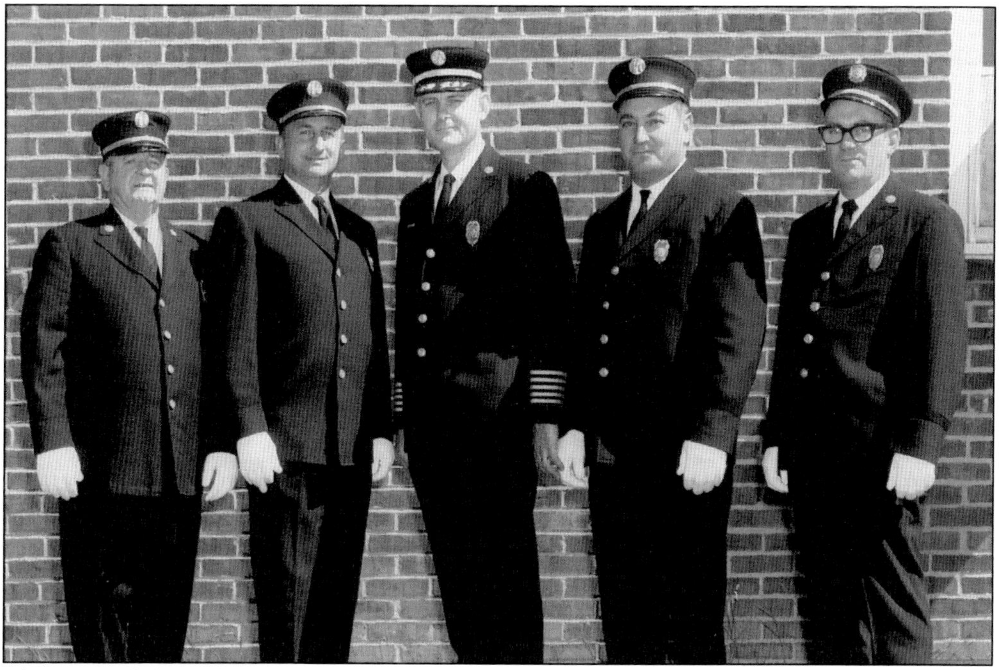

The original fire officers of the Edison Division of Fire pose alongside Station No. 1 with their newly appointed fire chief on Memorial Day 1970. Pictured from left to right are Capt. Edward Monaghan, Capt. Donald Dudics, Chief H. Ray Vliet, Capt. Milton Asprocolas, and Capt. Edward Costello. Dudics was promoted to the new position of deputy chief on May 9, 1973. (Courtesy of Donald Dudics.)

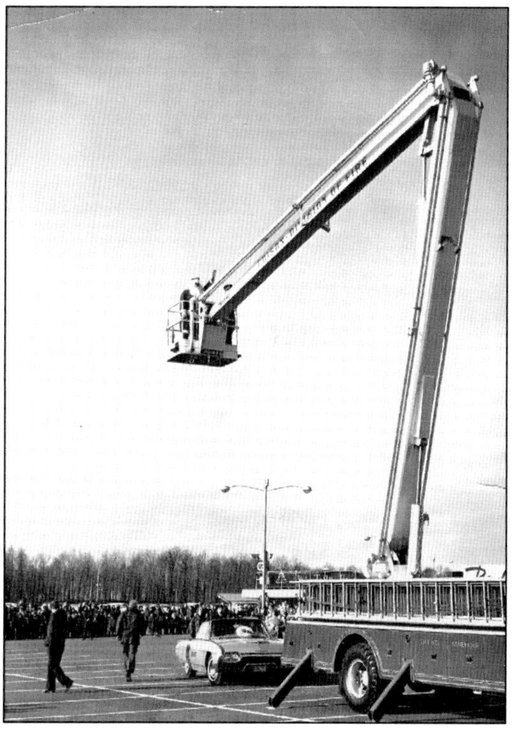

Edison firefighters are honored to elevate Santa Claus in the bucket of truck 1 at the reopening of the Menlo Park Mall in 1968. The mall, originally built as an open-air mall, was renovated to enclose the common area with a roof, which made it the first fully covered mall on the East Coast. (Courtesy of H. Ray Vliet.)

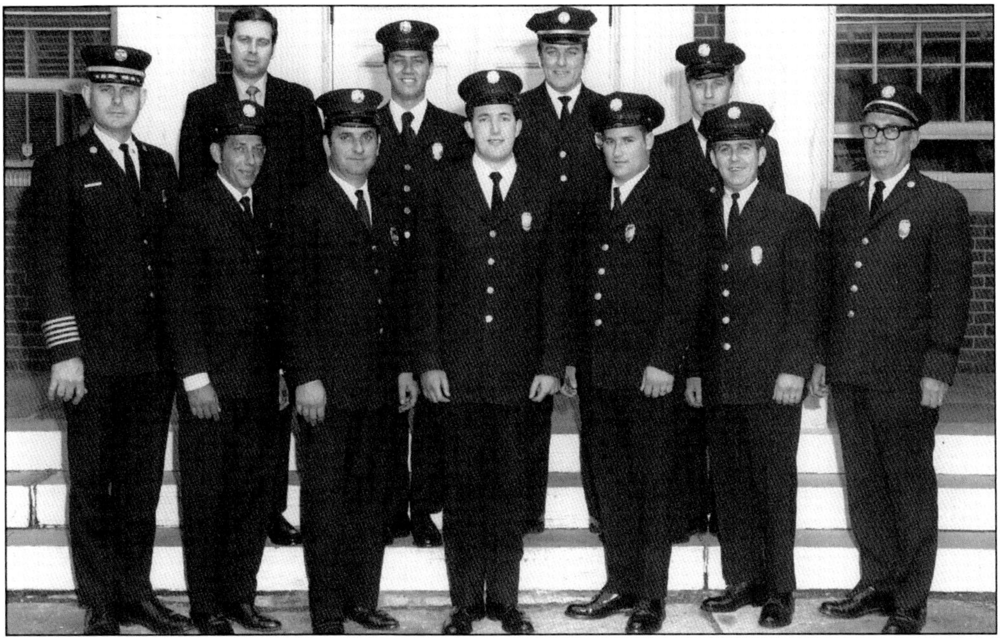

Chief H. Ray Vliet (far left) and Capt. Edward Costello (far right) pose in front of the old town hall with nine newly hired firefighters of the Edison Division of Fire on April 28, 1971. Standing next to Vliet from left to right are (first row) Kenneth Kuzma, John Spitler, G. Robert Campbell, James Montanye, and James Melynak; (second row) Ralph Banks, Robert Stramara, Herbert Eayres Sr., and Wesley Latham. (Courtesy of Herbert Eayres Jr.)

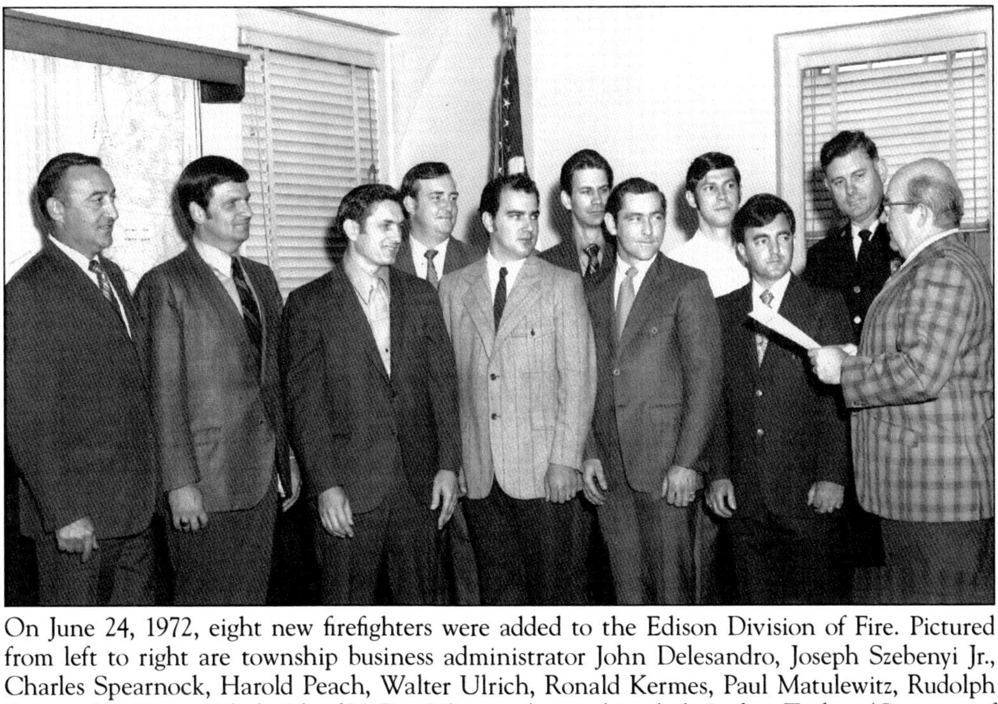

On June 24, 1972, eight new firefighters were added to the Edison Division of Fire. Pictured from left to right are township business administrator John Delesandro, Joseph Szebenyi Jr., Charles Spearnock, Harold Peach, Walter Ulrich, Ronald Kermes, Paul Matulewitz, Rudolph Asprocolas, Emery Tibok, Chief H. Ray Vliet, and township clerk Arthur Tucker. (Courtesy of Joseph Szebenyi Jr.)

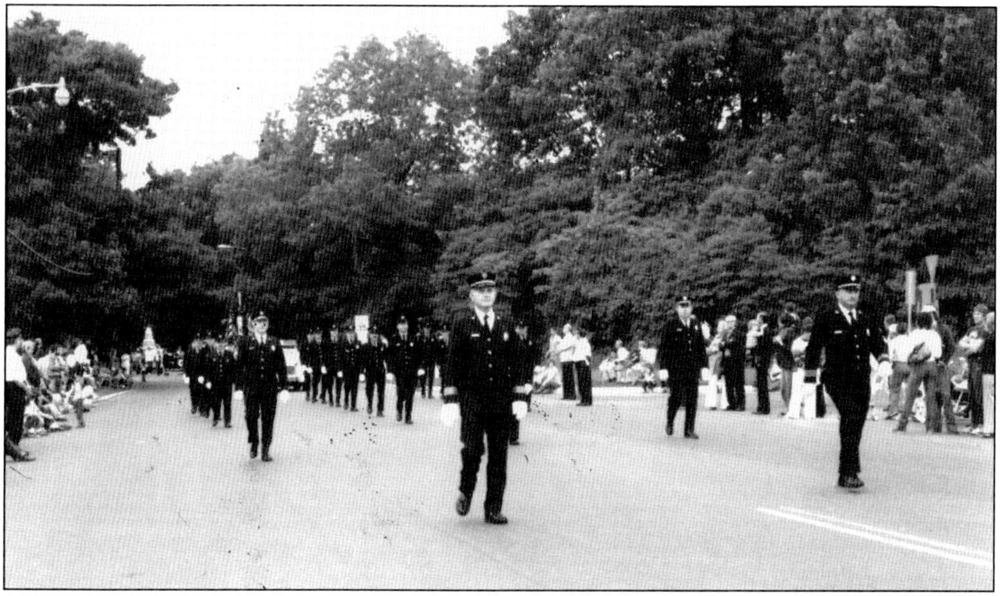

Chief H. Ray Vliet and newly appointed Deputy Chief Donald Dudics are leading the members of the Edison Division of Fire down Plainfield Avenue during the 1973 Memorial Day parade. (Courtesy of H. Ray Vliet.)

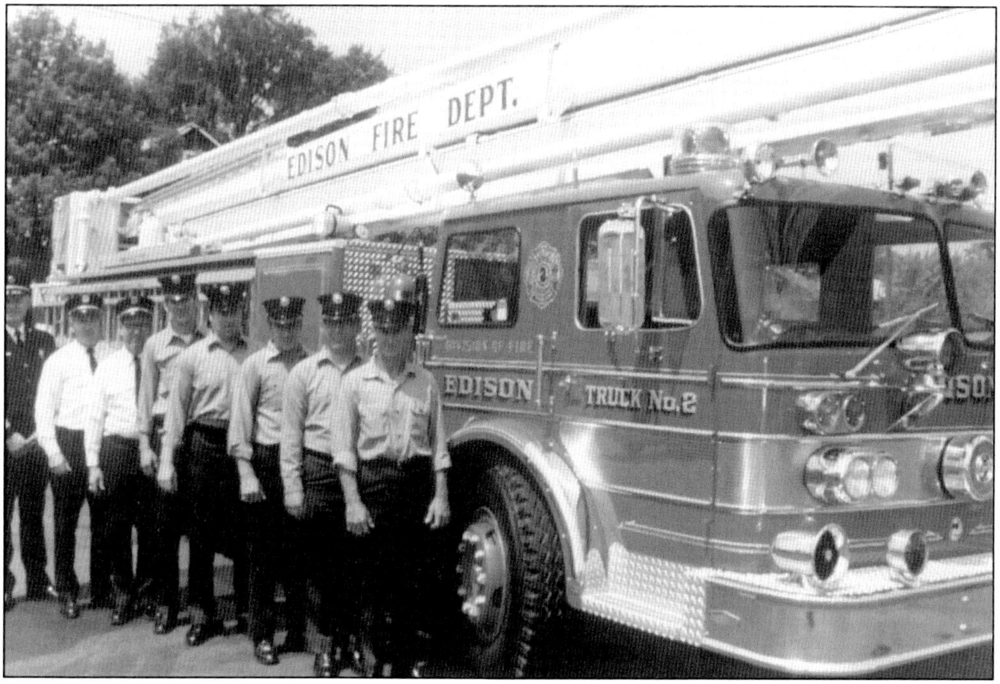

The Edison Division of Fire received its second aerial truck in July 1973. This 85-foot Oshkosh Snorkel, which was purchased for $100,000, was placed in service at Station No. 2 on Route 27 and the 1961 Ford Pitman Snorkel was moved to Fire Headquarters on Amboy Avenue. Pictured alongside truck 2 are (from back to front) Chief H. Ray Vliet, Capt. Richard Van Sickle, Capt. William Schnieder, firefighters Ronald Kermes, Charles Toth, Robert Yackel, Albert Lamkie, and Richard Latham. (Courtesy of Brian Latham.)

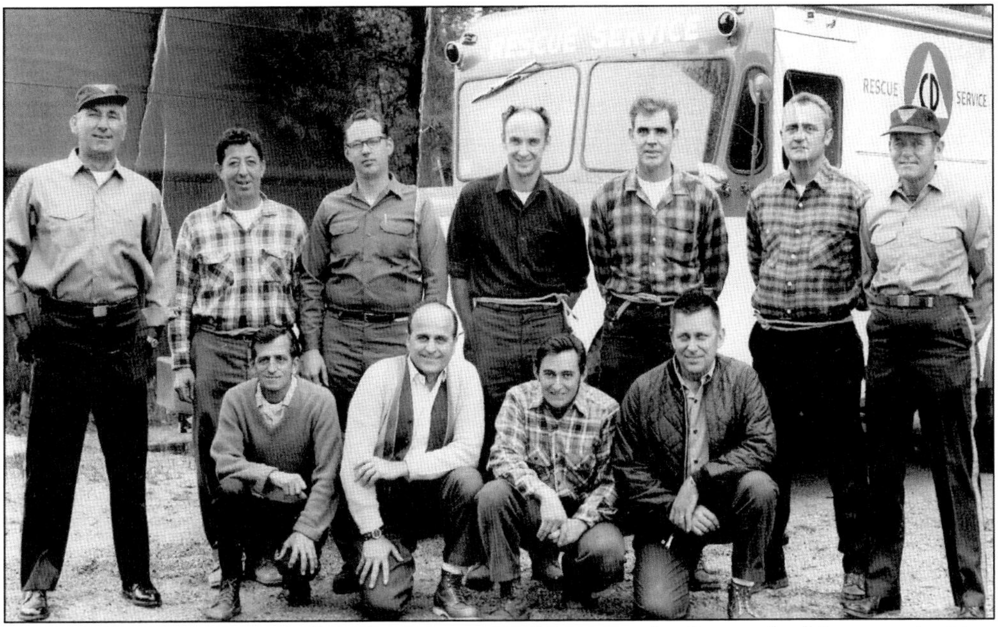

During the 1970s and 1980s, the Edison Division of Fire periodically sent firefighters to the New Jersey State Police Training Facility in Hammington to learn technical rescue procedures. The firefighters that attended the week-long course would stay and sleep in the barracks. This 1973 photograph shows four Edison firefighters, along with firefighters from other departments, just after the completion of the course. Kneeling in the first row, second from right, is Ralph Ambrosio Jr. with Joseph Horvath directly behind him, and Steven Mozsgae, kneeling far right with Robert Coleman standing directly behind him. (Courtesy of Ralph Ambrosio Jr.)

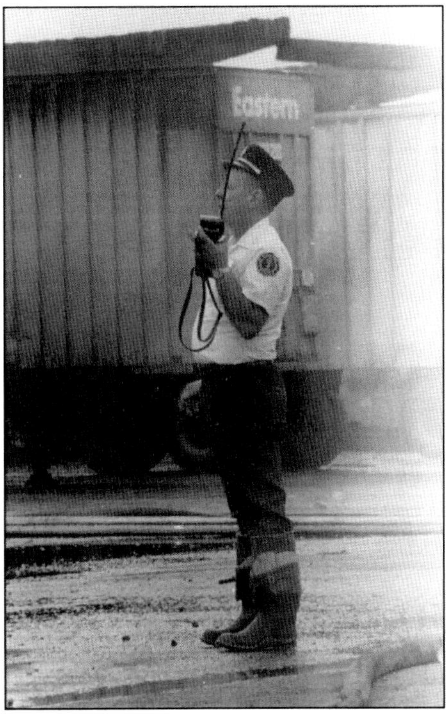

Capt. Robert Meluski takes command of a fire that ripped through Eastern Trucking on August 3, 1977. Meluski, an executive member of the firefighters union in the 1970s, pushed for the execution of a written contract. Prior to 1977, all agreements between career firefighters and Edison Township were verbally conducted and bonded by hand-shakes. (Courtesy of Thomas Walsh.)

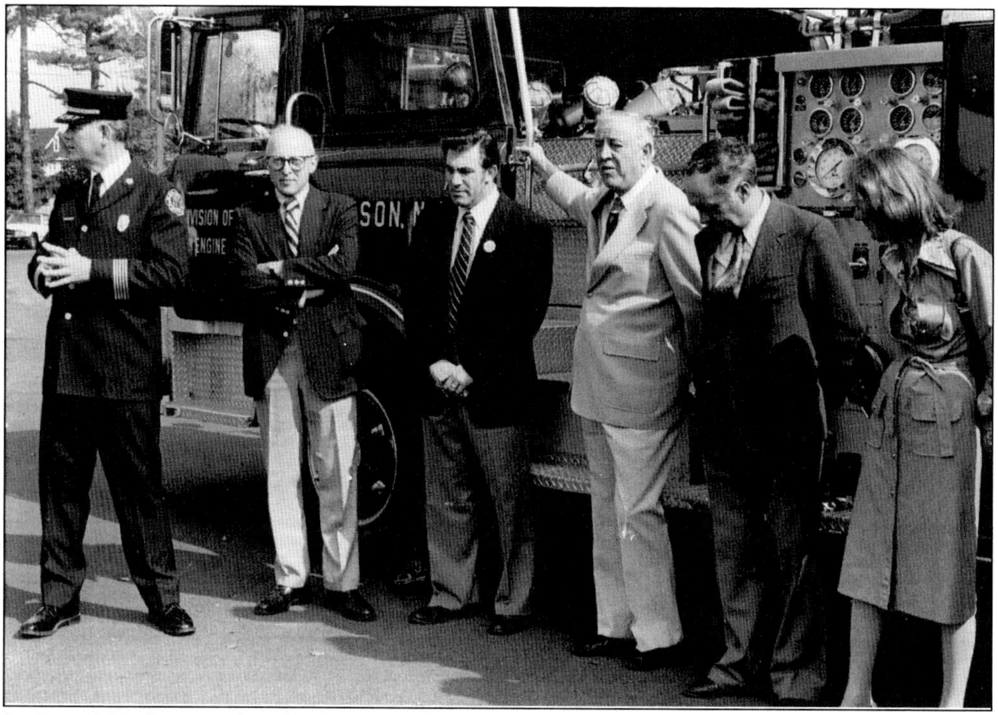

In this 1979 photograph, Chief H. Ray Vliet is showing council member Lewis Bloom, council member Thomas Paterniti, Mayor Anthony Yelencsics, council member John Hogan, and council member Margery Golin one of the new 1978 Mack fire engines. The engines were purchased to replace engine 6 and engine 12. They were powered by Mack Diesel Turbo engines with a 1,250-gallon-capacity pump. (Courtesy of H. Ray Vliet.)

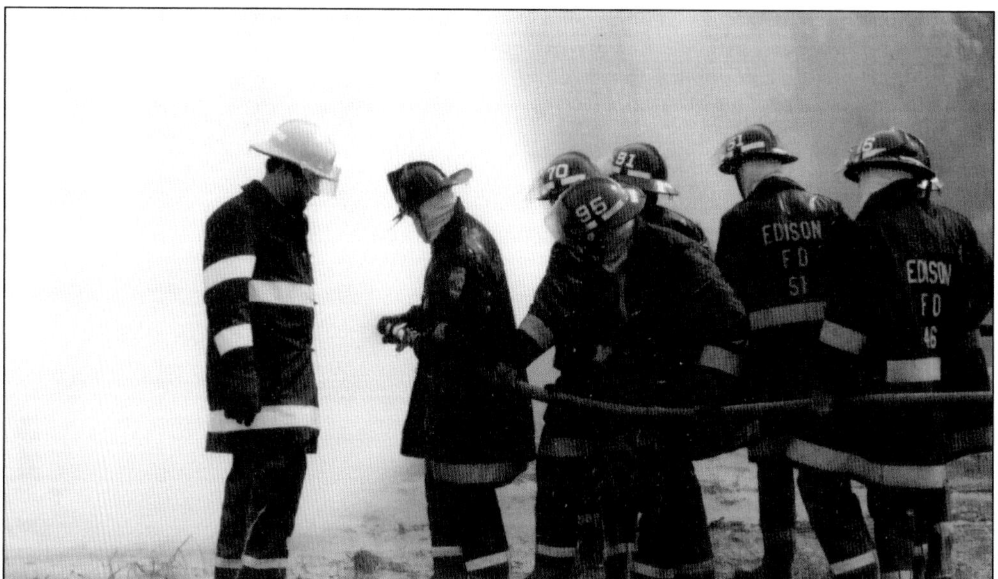

This photograph was taken at the Burlington Fire Academy for live fire training in the spring of 1982. (Courtesy of Brian Latham.)

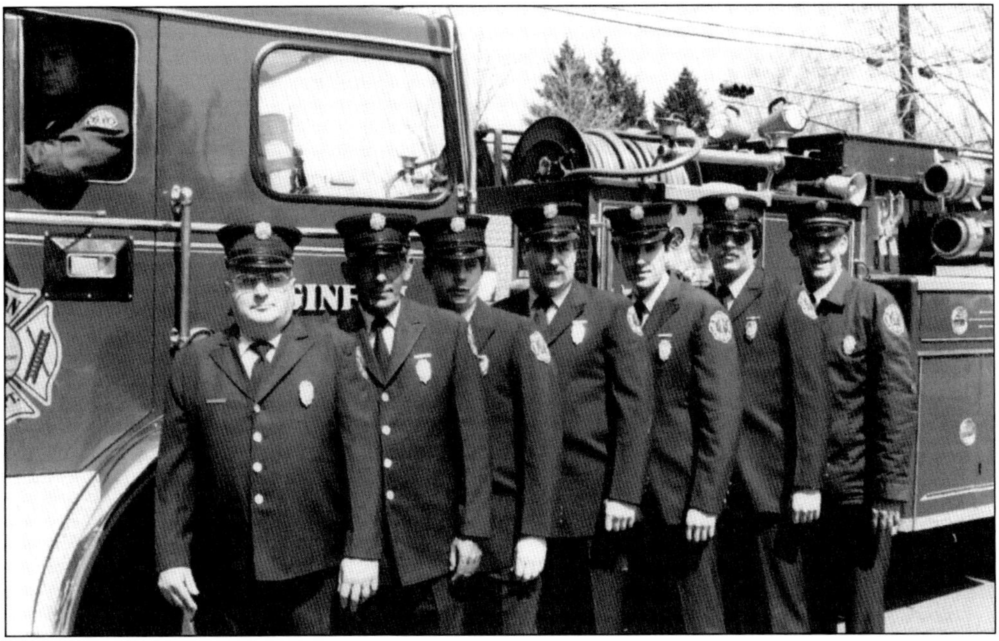

Firefighters pose alongside engine 1 after an award ceremony for four of the eight pictured above. From left to right are Julius Yuhas, Kenneth Kuzma, Dennis Phellan, Robert Yackel, Robert Horvath, Robert Sofield, Capt. John Lindquist, and Charles Toth sitting behind the wheel. On that day, four firefighters were recognized for their heroic actions at a Morse Avenue house fire that happened on October 27, 1982: Kuzma, Yackel, and Horvath received valor awards, and Sofield received a life-saving award. (Courtesy of Robert Sofield.)

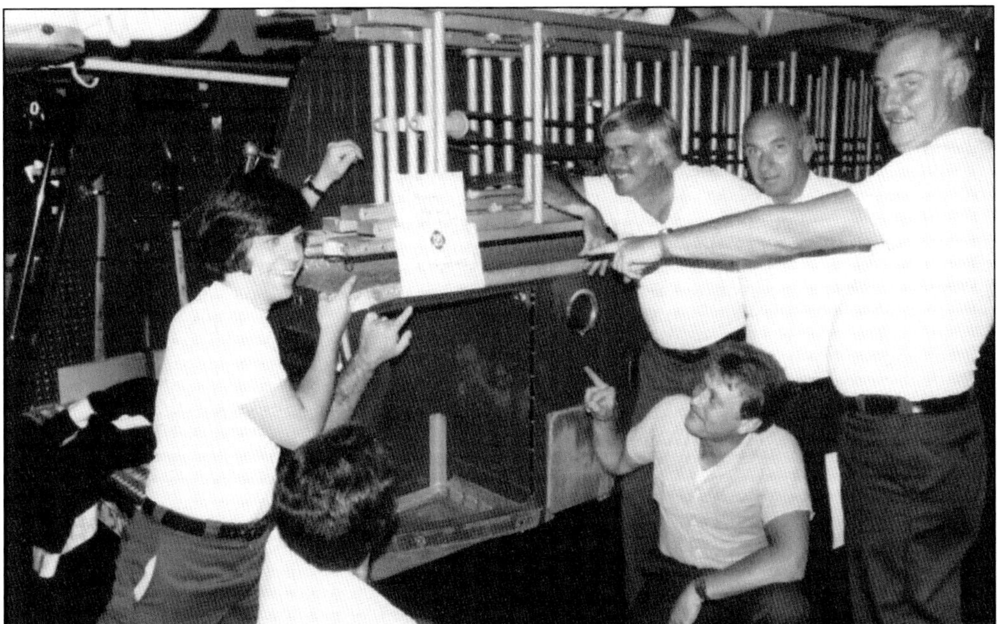

This 1983 photograph was taken shortly after a member of the Edison Division of Fire accidentally damaged truck 1's rear compartment door. As expected, fellow firefighters from D-Shift had fun pointing out the damage. (Courtesy of Donald Freeman.)

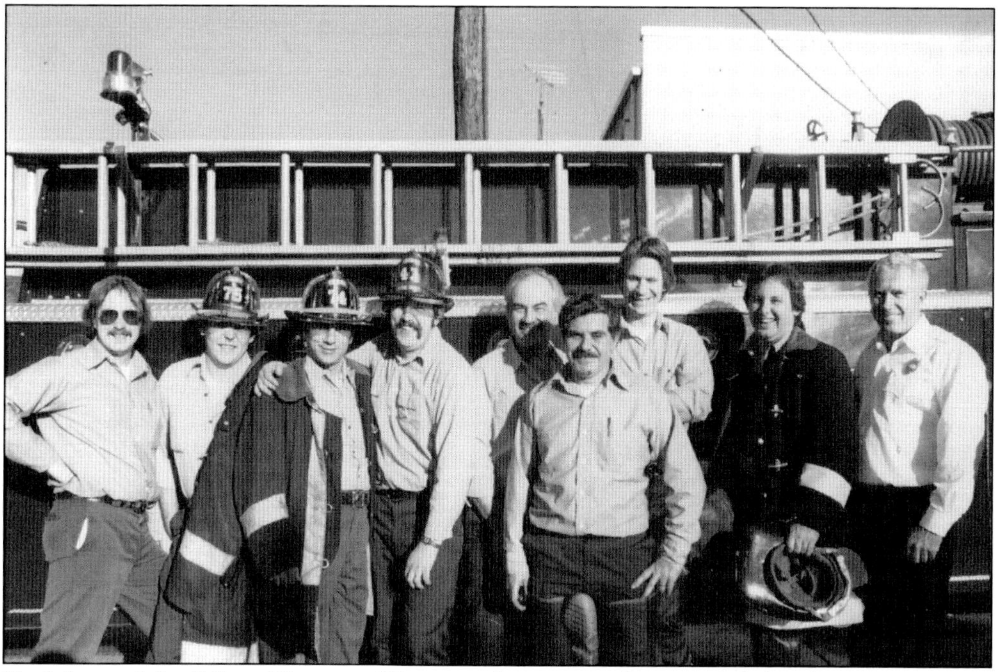

Firefighters of the Edison Division of Fire pose for a group photograph behind Station No. 2 in the early 1980s. Pictured from left to right are Jay Marino, Eugene Whitecavage, Peter Anselmo, G. Robert Campbell, John Dublanyk, Anthony Colletto, Ronald Kermes, Jeff Voorhees, and Capt. Richard Van Sickle. (Courtesy of Eugene Whitecavage.)

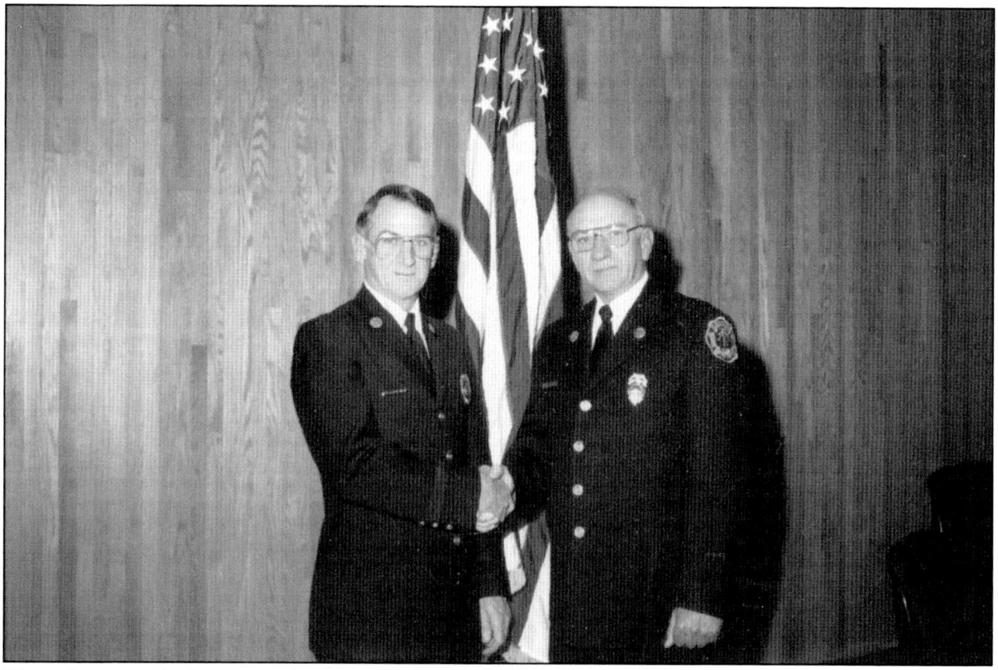

This photograph shows newly appointed chief Richard Latham (left) and Deputy Chief Arthur Harmon shaking hands during their promotion ceremony at council chambers on January 1, 1985. (Courtesy of Richard Latham.)

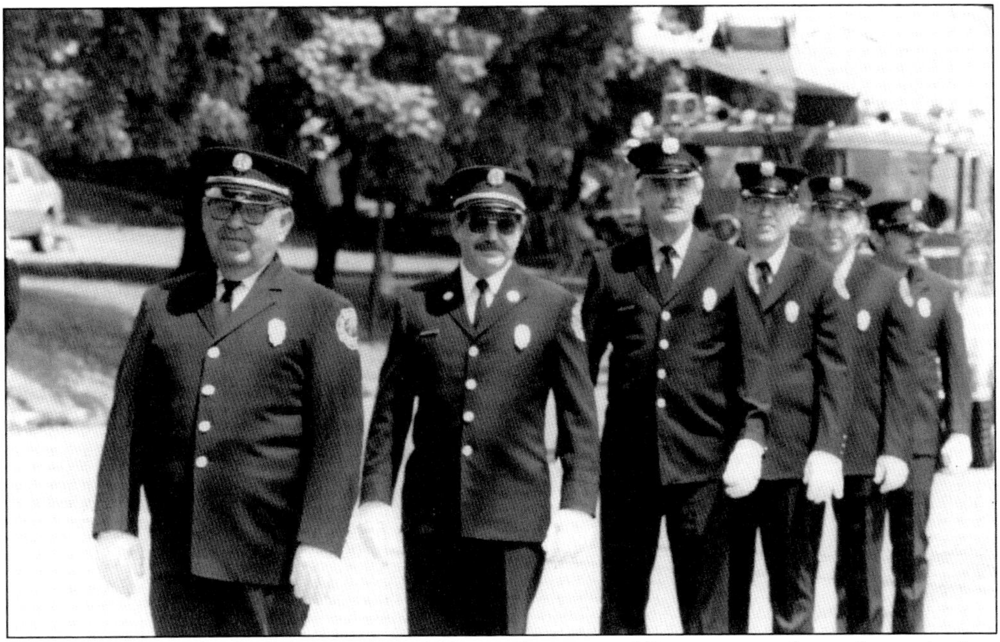

In this photograph, firefighters of the Edison Division of Fire are marching down Amboy Avenue in the 1984 Memorial Day parade. In the front are Captains Julius Yuhas and Anthony Colletto, followed by firefighters Donald Freeman, Albert Lamkie, Edward Guarnieri, and Ralph Ambrosio. (Courtesy of Brian Latham.)

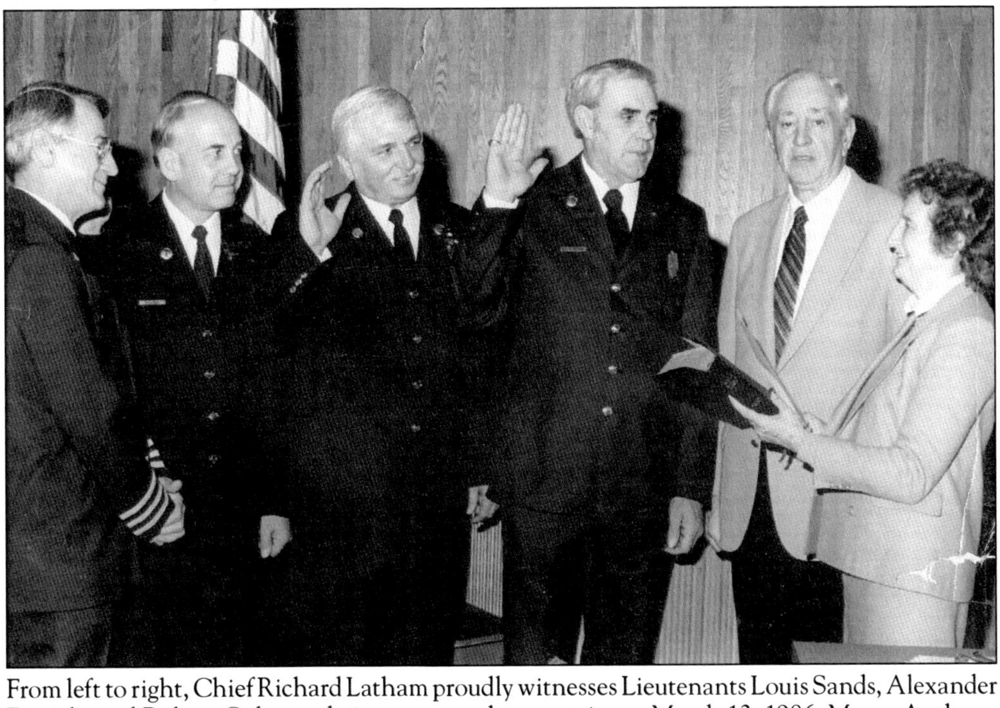

From left to right, Chief Richard Latham proudly witnesses Lieutenants Louis Sands, Alexander Danish, and Robert Coleman being promoted to captain on March 13, 1986. Mayor Anthony Yelencsics and township clerk Lucille Tucker conduct the official procedure. (Courtesy of Robert Coleman.)

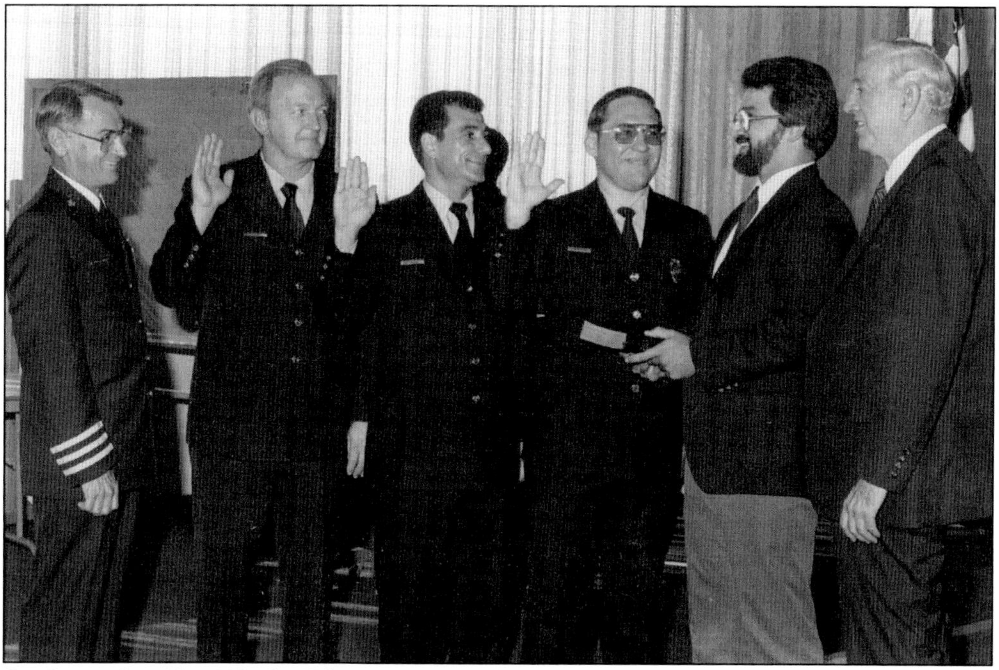

On September 25, 1985, the Edison Division of Fire promoted three firefighters and hired five. Shown above, firefighters Carl Woods (left) and Peter Anselmo (middle) are being promoted to fire inspectors while firefighter Albert Lamkie (right) is being promoted to captain. Shown below from left to right, Joseph Szebenyi III, Albert Stevens, Brian Latham, Brian Daugherty, and Mark Anacker are being sworn in as career firefighters. In both photographs, business administrator Larry Pollex is holding the Bible as Chief Richard Latham (left) and Mayor Anthony Yelencsics observe the official ceremony. (Courtesy of Brian Daugherty.)

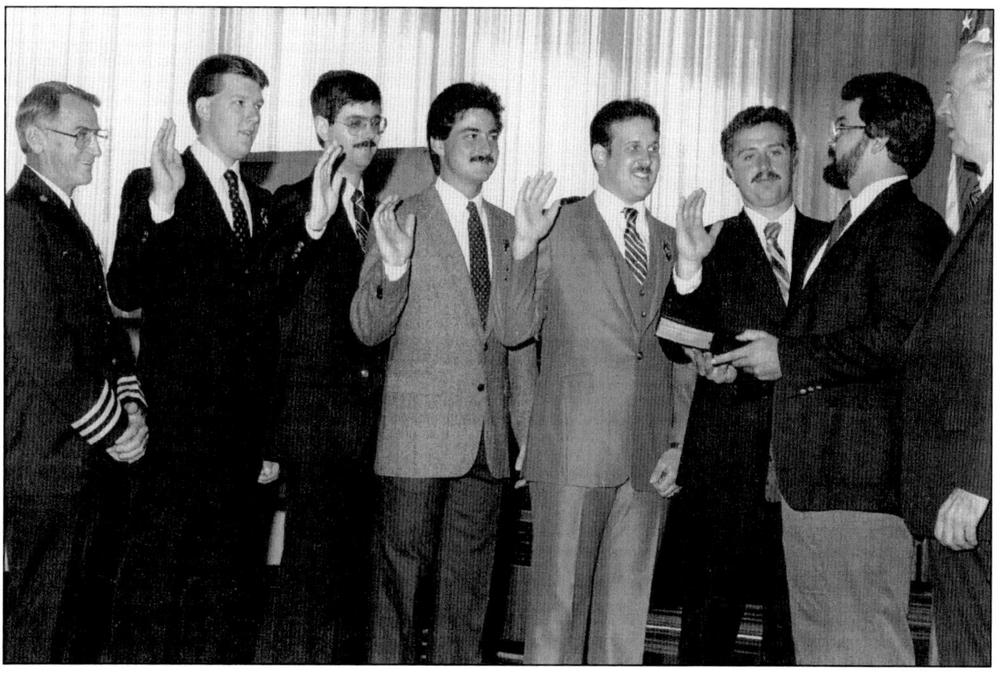

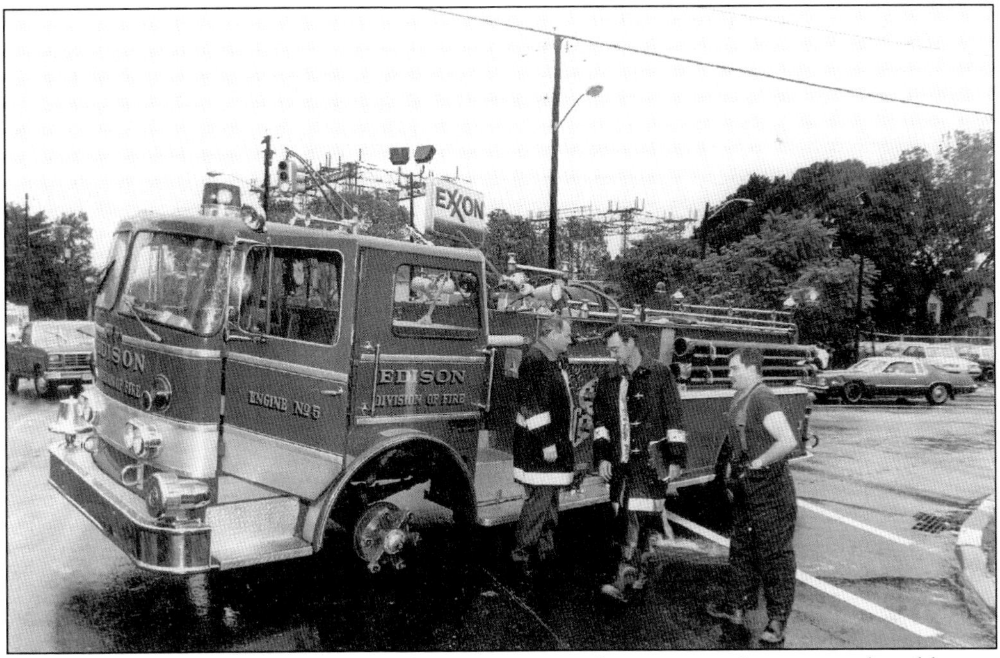

While responding to an alarm, firefighters on engine 5 encountered an unusual problem at the corner of Amboy Avenue and Lake Avenue in Metuchen. Firefighter Leon Mazur (left), Lt. Joseph Horvath (center), and volunteer Ronald Lorrentiev assess the embarrassing situation. (Courtesy of Jim D'Amico.)

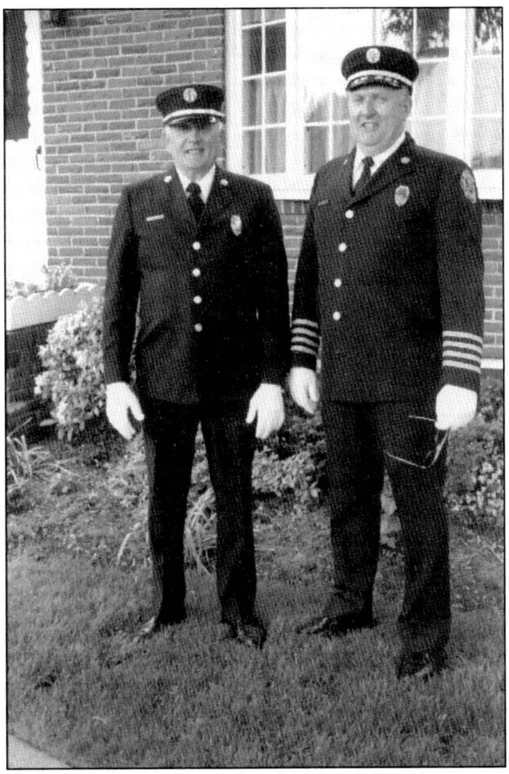

Capt. Charles Grand-Jean Jr. (right) was promoted to deputy chief on January 1, 1986. This photograph of the newly appointed deputy chief was taken on Memorial Day 1986 with his brother Robert, a retired captain. (Courtesy of Robert Grand-Jean.)

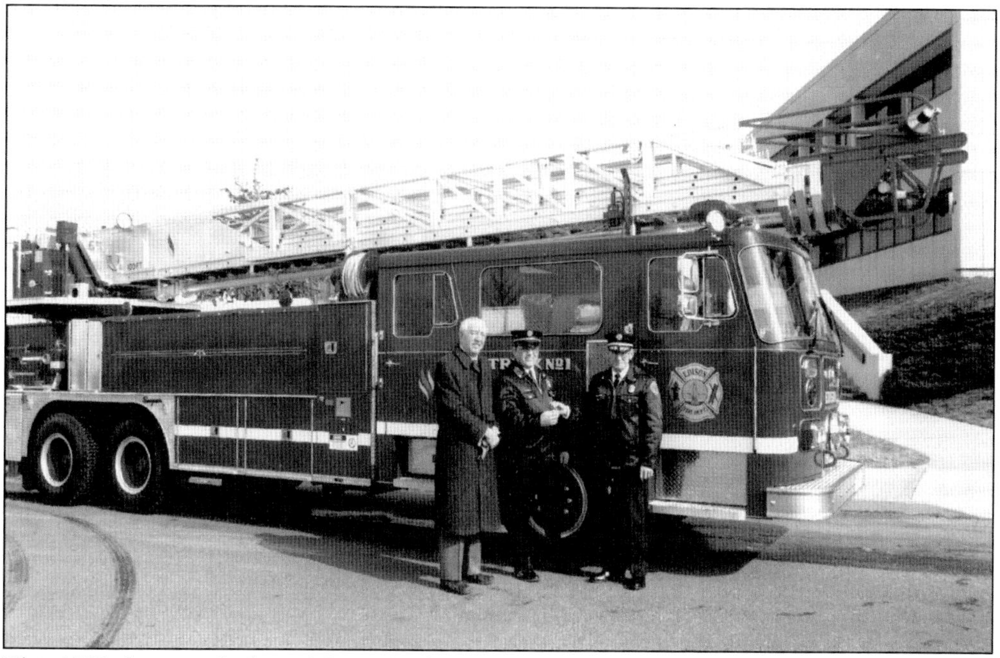

This 1987 photograph shows Mayor Anthony Yelencsics, training captain Albert Lamkie, and Chief Richard Latham posing in front of the new Seagrave aerial truck. This truck, which was equipped with a 100-foot aerial ladder, replaced the 85-foot Oshkosh Snorkel that was housed at Station No. 2 on Route 27. (Courtesy of Brian Latham.)

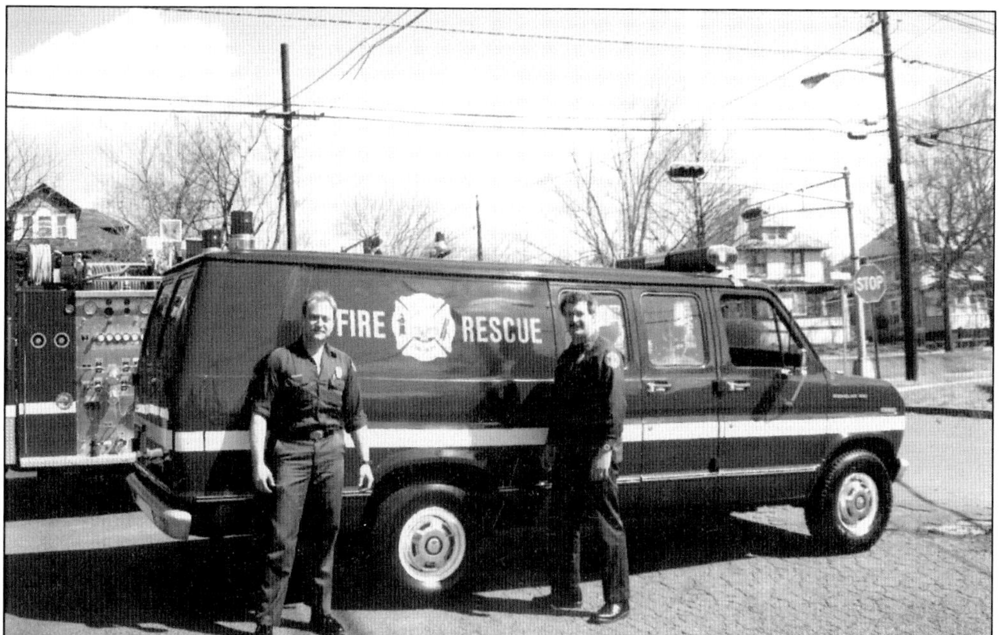

Firefighter Robert LaCour (left) and firefighter Donald Gunthner pose in front of the first fire rescue in November 1989. This Ford Econoline van, which was equipped with basic fire and first-aid gear, was placed into service to provide special firefighting services and emergency medical care. (Courtesy of Robert LaCour.)

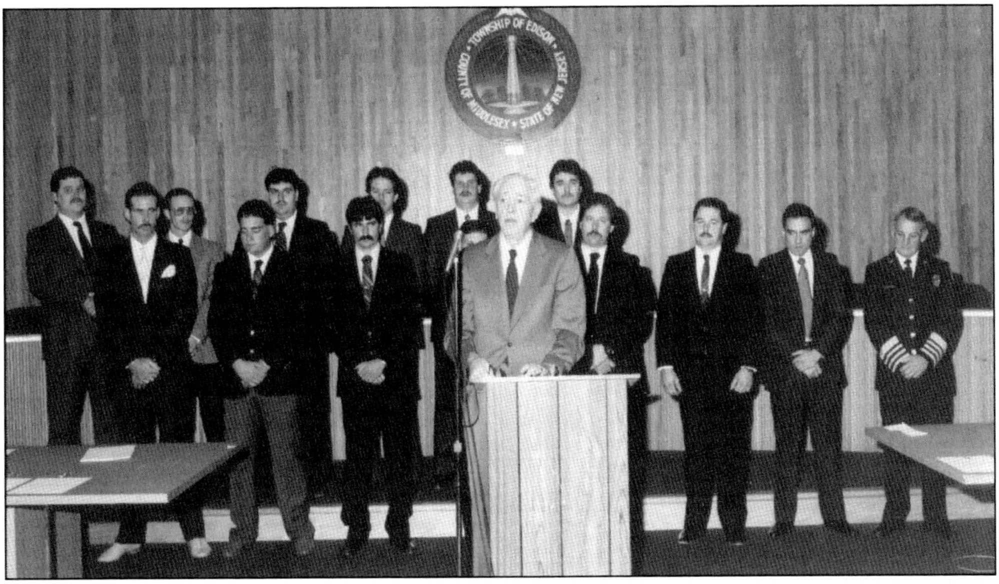

Mayor Anthony Yelencsics, standing at the podium, welcomes 13 new firefighters to the Edison Division of Fire on November 2, 1988. Pictured from left to right are (first row) Eric Johnson, Anthony Landi III, Scott McCloughan, Kenneth Milcsik, Gregory Schaefer, Michael Yock, Thomas Shjarback, and Chief Richard Latham; (second row) Gregg Alleman, Richard Aszman, Charles Copperthwaite, Thomas Droppa, Louis Gurdon Jr., and Gregory Gush. This was the last group to be sworn in by Yelencsics. (Courtesy of Scott McCloughan.)

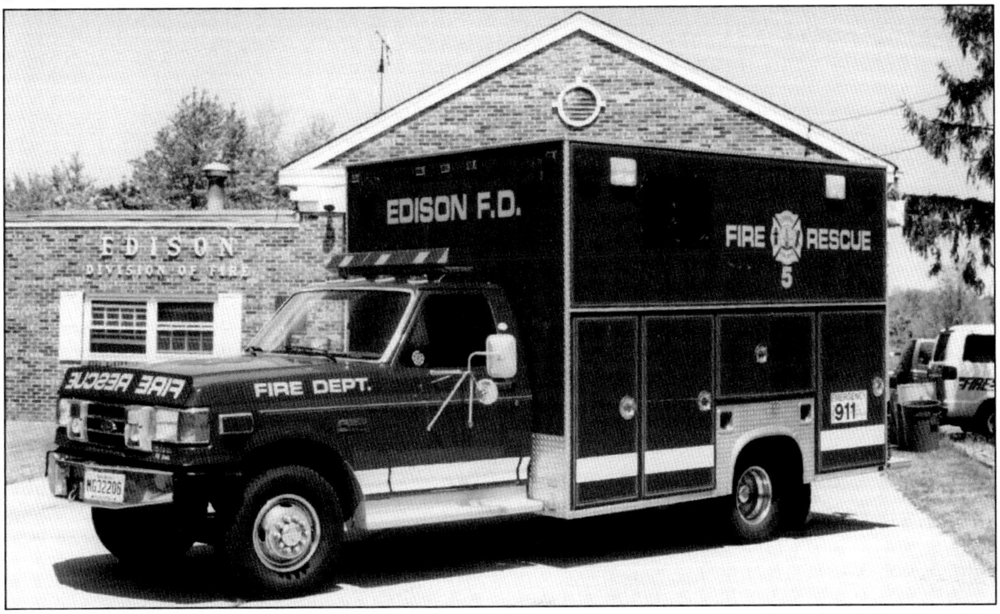

In 1994, fire rescue was upgraded from the Ford Econoline van to this 1991 Ford F-350 utility body. In addition to the basic firefighting and emergency medical equipment, a set of pneumatic extrication tools were carried on board. All firefighters that rode this rig were required to be New Jersey State certified emergency medical technicians. This photograph was taken in 1997 when the unit was moved to Station No. 4 and became Fire Rescue No. 5 to cover the northern section of Edison. (Courtesy of Edison Volunteer Fire Company No. 1, photograph by John Calderone.)

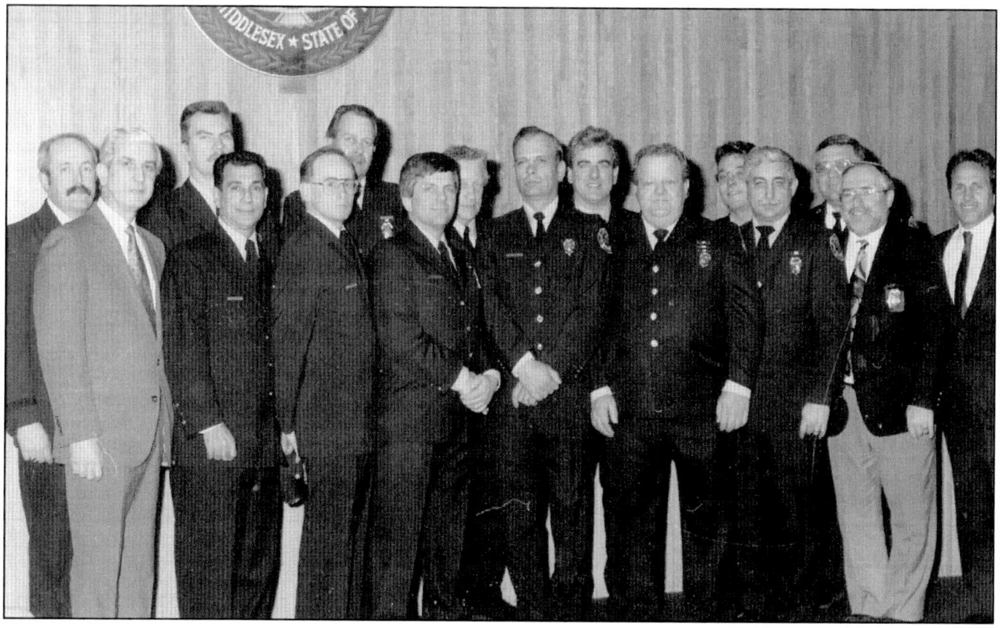

In May 1994, eleven firefighters were promoted to the rank of lieutenant. Pictured from left to right are (first row) Michael Pucchio (public safety director), Peter Anselmo, Michael Costello, Norman Jensen, Peter Borwegen, Keith McGorvin, Patrick Novia, Rabbi Rosenberg (township chaplain), and Mayor George Spadoro; (second row) Deputy Chief G. Robert Campbell, Edward Koehler, Ralph Banks, Richard Campbell, Walter Ulrich, Eugene Whitecavage, and Chief Albert Lamkie. (Courtesy of Richard Campbell.)

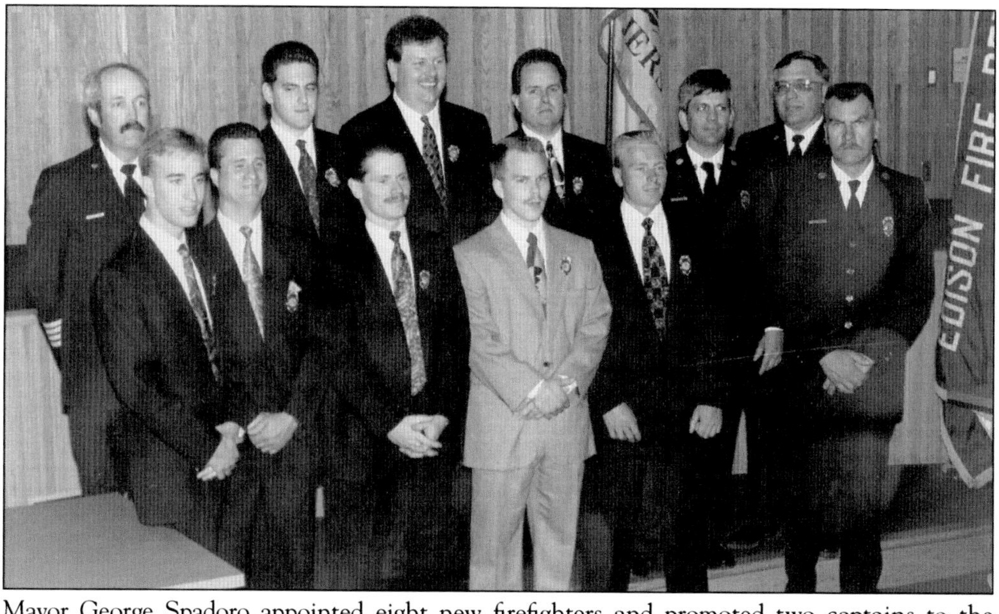

Mayor George Spadoro appointed eight new firefighters and promoted two captains to the Edison Division of Fire on May 22, 1995. Pictured from left to right are Deputy Chief G. Robert Campbell, Eugene Enfield Jr., Francis Hohwald, John Palmer, Robert McDonnell, Richard Ball, James Gross, Daniel Wood, Michael Wheeler, Norman Jensen (captain), Chief Albert Lamkie, and Edward Kohler (captain). (Author's collection.)

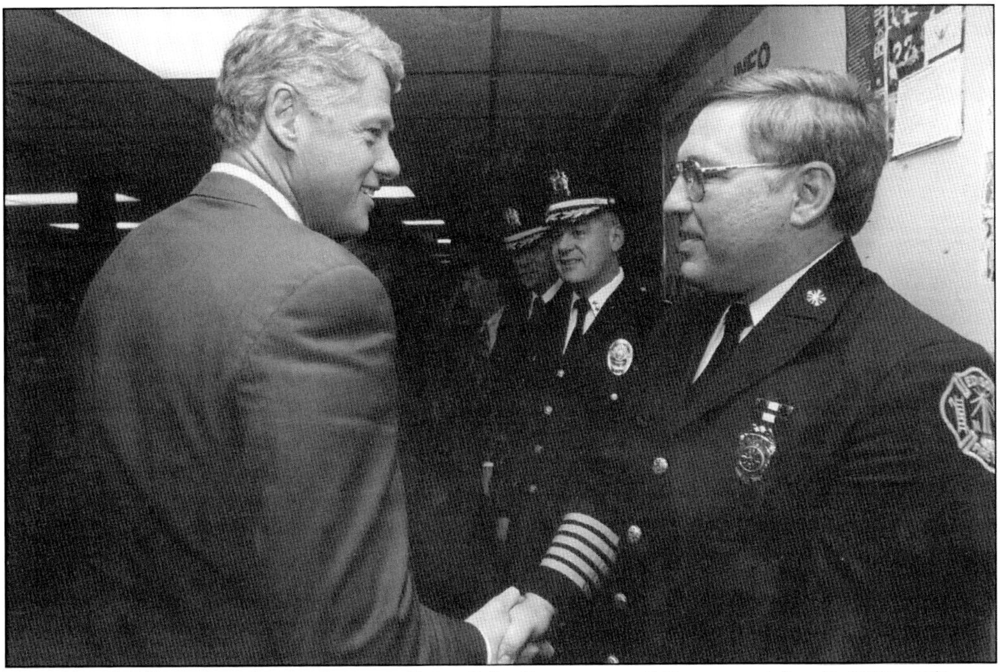

Fire Chief Albert Lamkie welcomes Pres. Bill Clinton to Edison on May 7, 1996. Clinton delivered a campaign speech at Middlesex County College while preparing for his next term in office. (Courtesy of Albert Lamkie.)

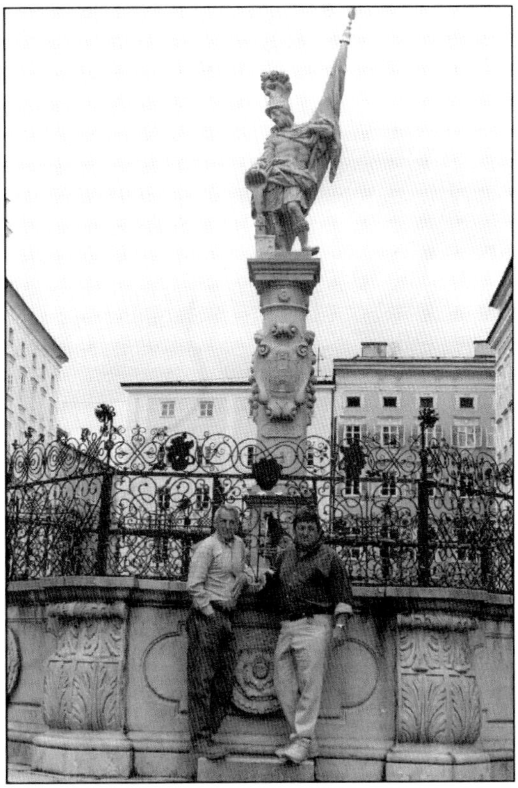

This 1996 photograph shows retired chief Richard Latham (left) and Deputy Chief Ralph Ambrosio Jr. posing in front of the St. Florian Monument in Salzburg, Austria. St. Florian, the patron saint for firefighters, was an officer in the Roman army during the third century. According to legend, St. Florian miraculously stopped an entire town from burning by throwing a single bucket of water onto the fire. (Courtesy of Richard Latham.)

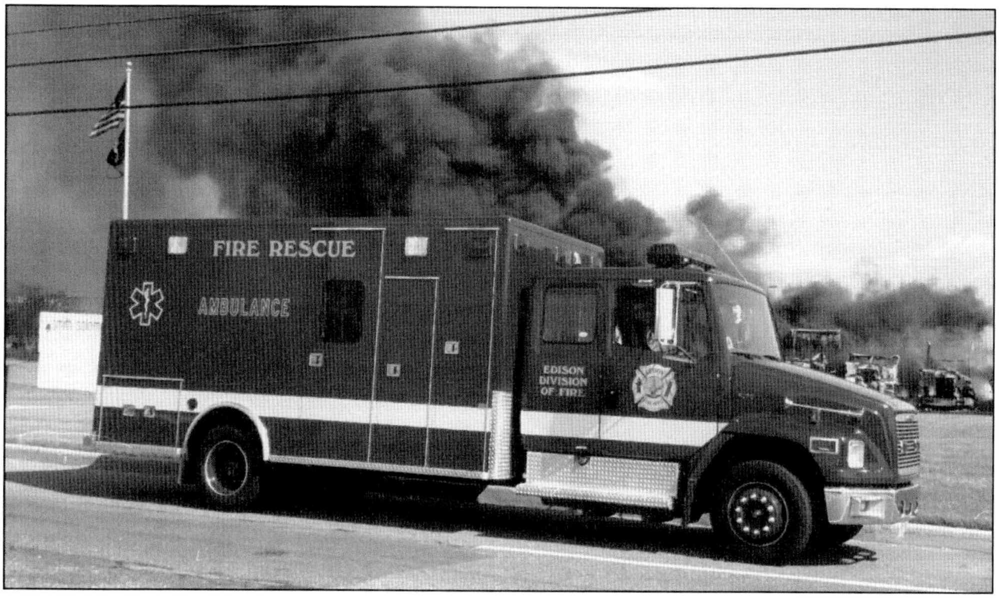

In 1997, the Edison Division of Fire purchased a new fire rescue. This specialized fire apparatus was not only capable of providing basic fire and rescue functions, but was also equipped with a stretcher and emergency medical supplies for transporting patients to local hospitals. This 1997 International Freightliner responded to various emergencies throughout Edison and racked up over 200,000 miles within eight years. It has been recently replaced by a 2004 Freightliner and has been designated as a reserve piece. (Courtesy of Edison Division of Fire, photograph by John Calderone.)

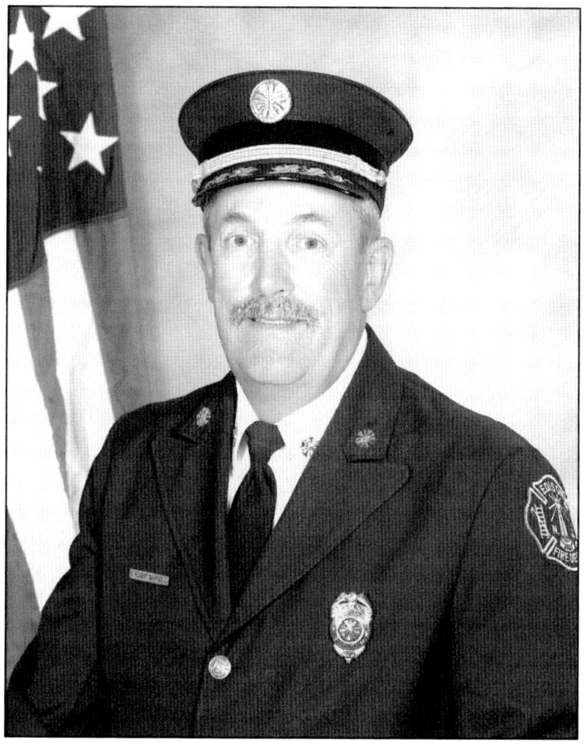

Deputy Chief G. Robert Campbell was appointed chief of the Edison Division of Fire by Mayor George Spadoro on January 6, 1999. Campbell, who is a lifelong resident of Edison, earned a bachelor's degree from Clemson University and master's degree from Rutgers University. (Courtesy of P.M. Studios.)

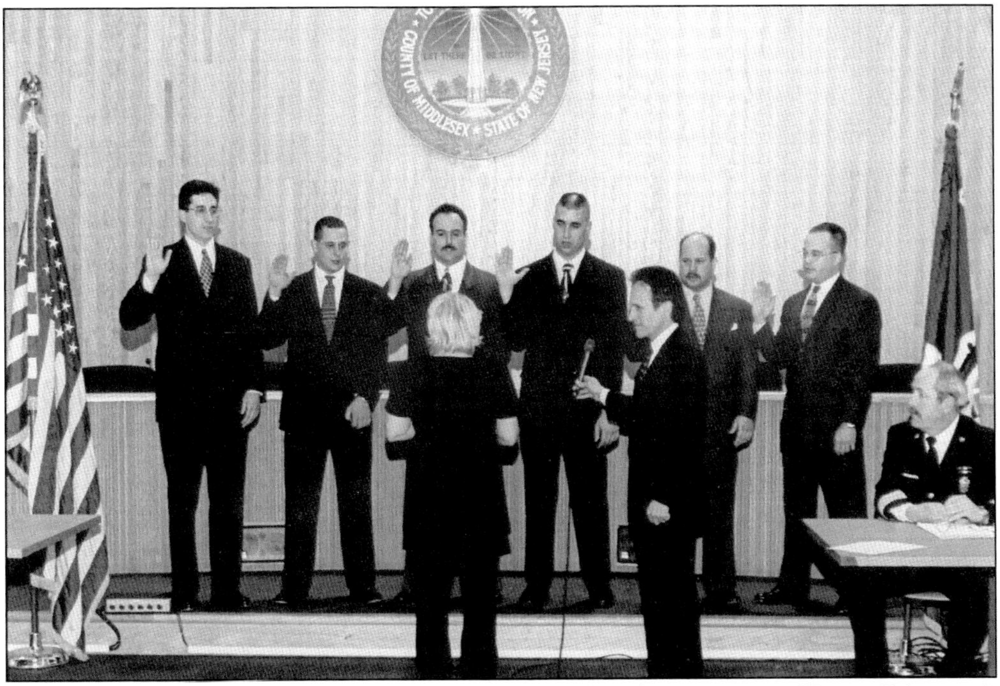

Mayor George Spardoro appointed six new firefighters to the Edison Division of Fire on February 3, 2000. James Walsh, Joseph Anselmo, Michael Sacchi, Patrick J. Novia, Samuel Kirkpatrick, and Kevin Hopkins each has his right hand raised as township clerk Renee Murphy officially swears the group in while Chief G. Robert Campbell looks on. (Courtesy of Patrick J. Novia.)

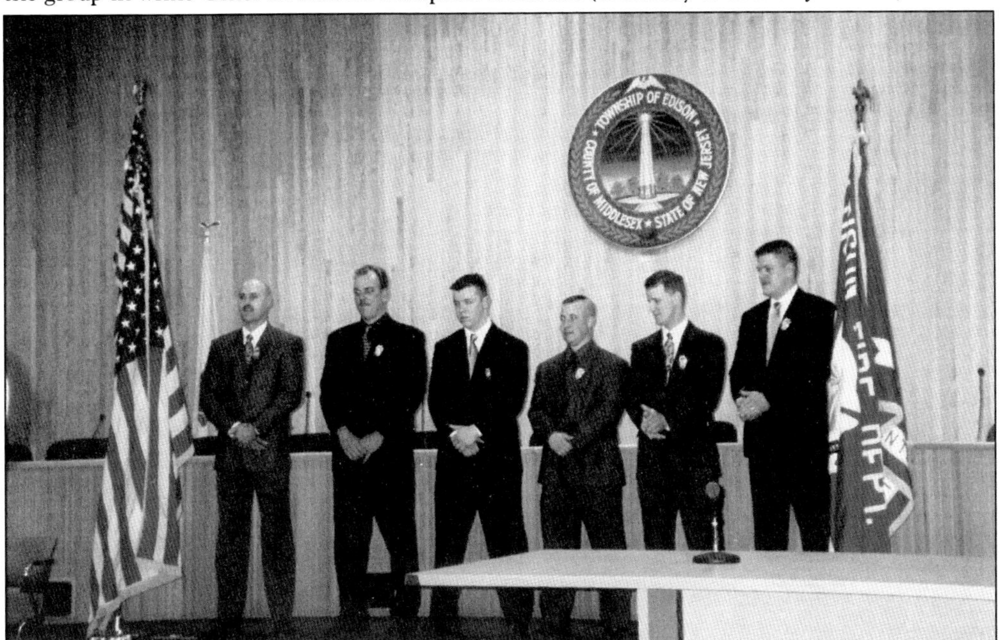

Standing at the front of council chambers are, from left to right, Paul Toth, Wayne Enoch, Michael Maurath, William Pellegrino, Allan Yourstone Jr., and Herbert Earyes Jr. as they are sworn in as career firefighters on March 23, 2001. (Courtesy of Wayne Enoch.)

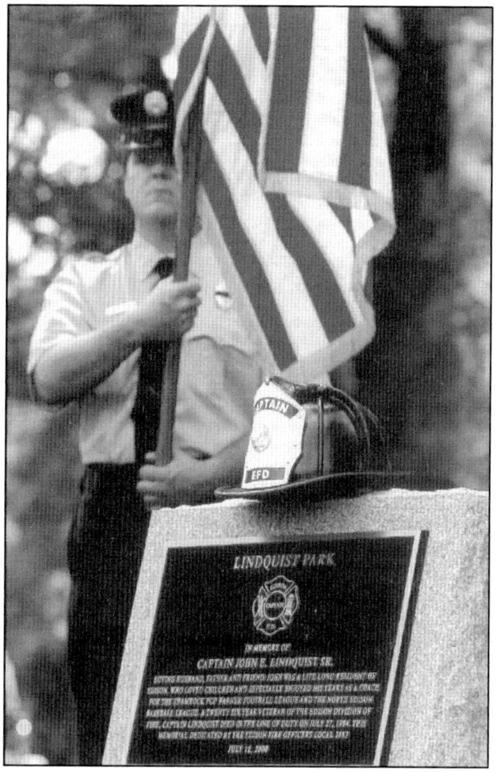

Capt. John Evans Lindquist, who died in the line of duty on July 27, 1984, while fighting a house fire, will always be honored and remembered for his dedicated service. Pictured at left is Edison firefighter Anthony Ambrosio III holding the American flag during the Lindquist monument dedication on July 25, 2004. The monument, which is located at Lindquist Park, was dedicated by the Edison Fire Officers Local No. 2883. (Courtesy Greater Media Newspapers, photograph by Chris Kelly.)

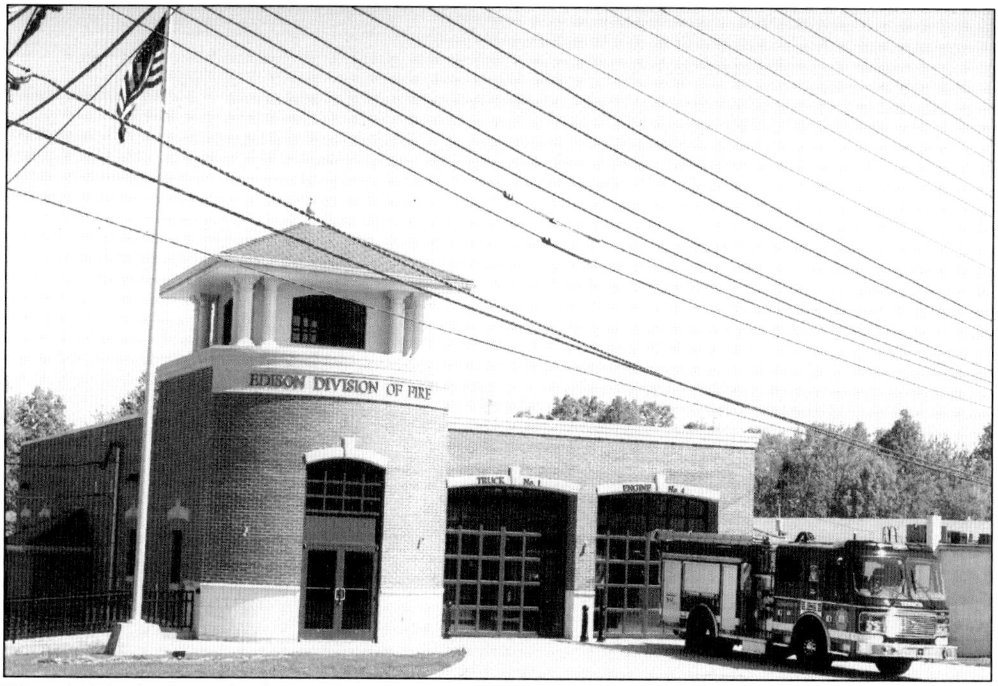

This 2006 photograph shows the newly renovated Station No. 2, located at 1997 Lincoln Highway (Route 27). The renovation project took two years and was completed by the Ferreira Group Product Managers. (Courtesy of Karen Daugherty.)

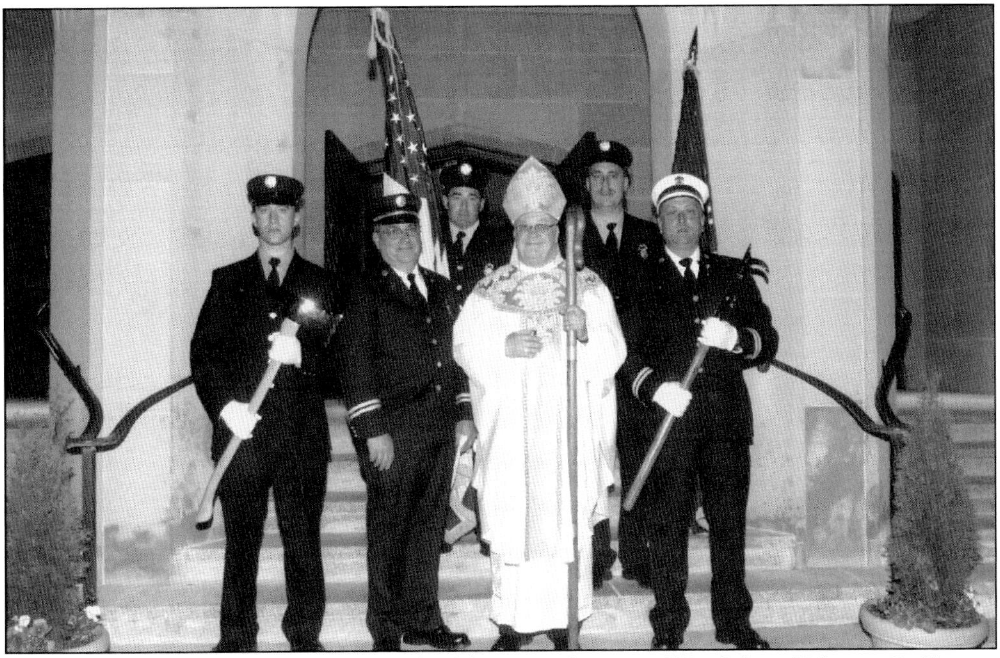

Bishop Paul Bootkoski was the principal celebrant at the annual Firefighters' Mass at St. Francis of Assisi Cathedral on April 21, 2007. After the tragic events of September 11, 2001, the Diocese of Metuchen created a special mass held annually to honor and bless all firefighters. Firefighters Eugene Enfield Jr., Capt. David Davis, Robert Mellinger, Patrick Novia, and Capt. Robert LaCour (honor guard originator) were honored to be photographed with Bishop Bootkoski. (Courtesy of Kris Rupp Photography.)

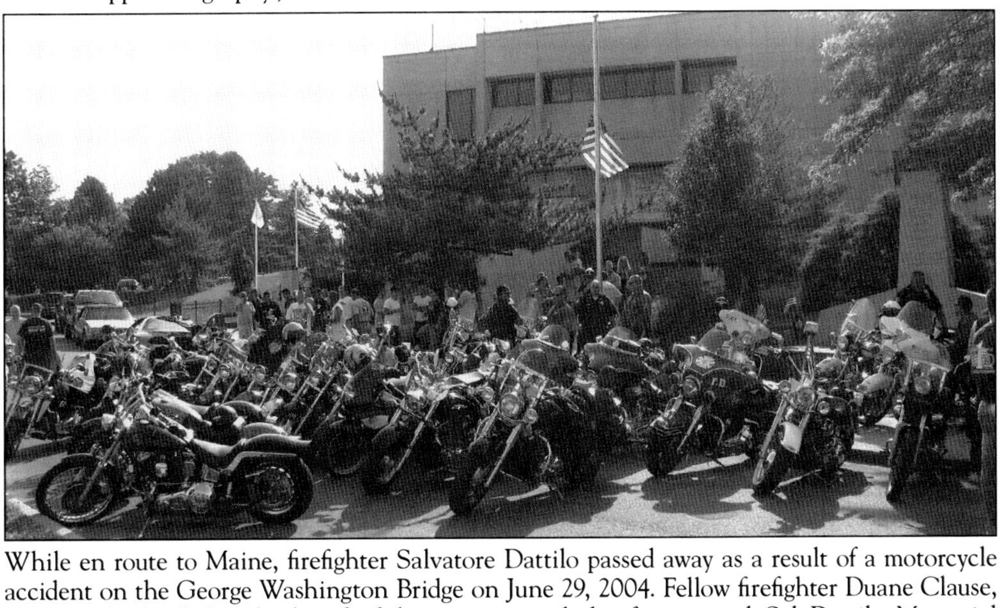

While en route to Maine, firefighter Salvatore Dattilo passed away as a result of a motorcycle accident on the George Washington Bridge on June 29, 2004. Fellow firefighter Duane Clause, along with the help of other firefighters, organized the first annual Sal Dattilo Memorial Motorcycle Ride on July 8, 2007. This photograph was taken in front of the Firefighters' Memorial at the Edison Municipal Complex just before the motorcyclists departed for the benefit ride. (Courtesy of Glenn Osborne.)

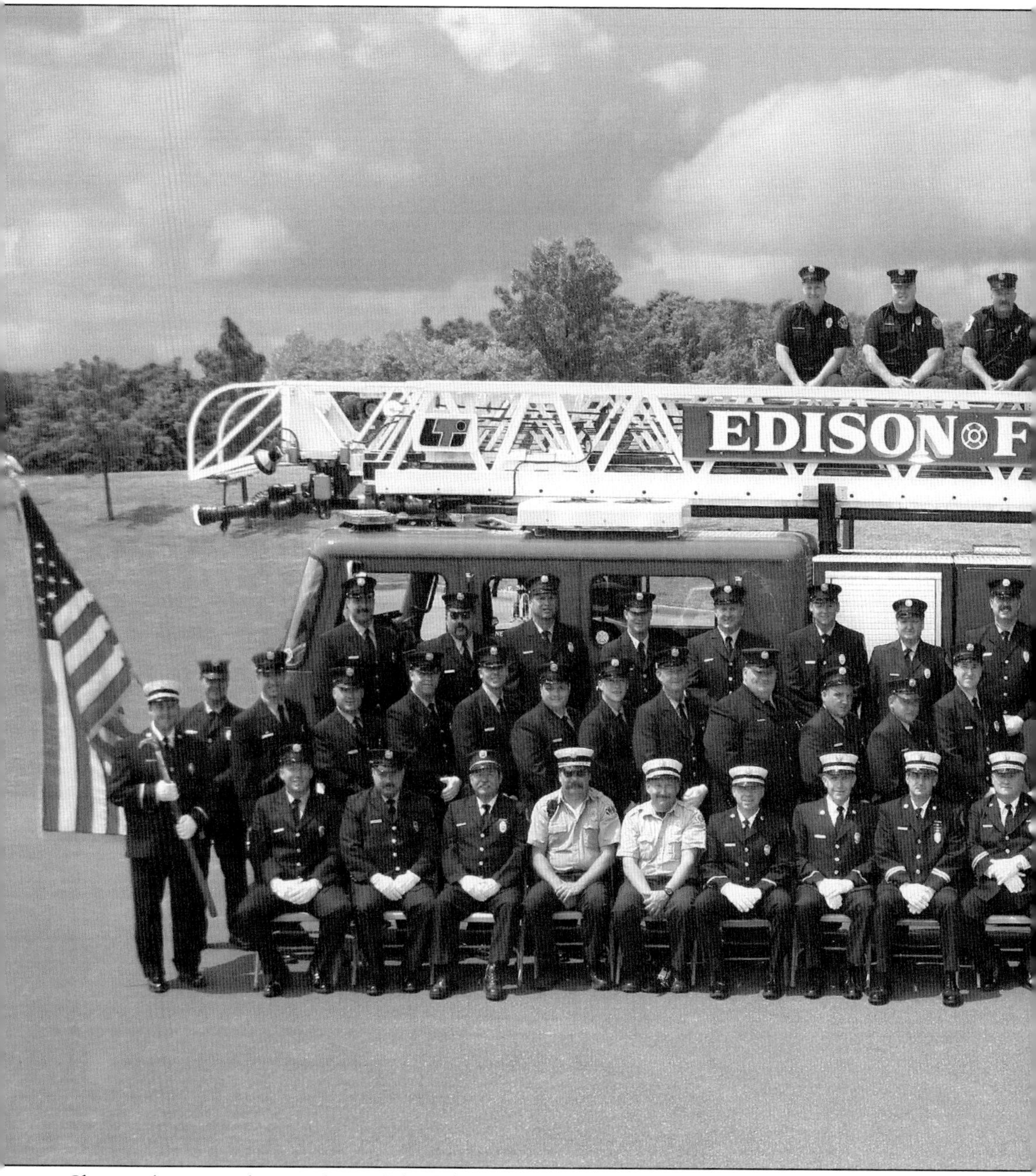

Shown above are the members of the Edison Division of Fire, who were in attendance for the firefighters' Memorial Day service at the Edison Municipal Complex on May 27, 2007. Pictured from left to right are (first row) firefighters Michael Banbor, Samuel Kirkpatrick, Inspector William Enoch, Lt. Bruce Weigel, Lt. Kenneth Milcsik, Lt. Joseph Gaul, Lt. Scott Boland, Capt. Brian Latham, Capt. James Montanye, acting chief Norman Jensen, Capt. David Davis, Capt. Robert LaCour, Lt. Joseph Szebenyi III, Lt. David Milcsik, firefighters Henry Tempe, Scott Law, Herbert Eayres Jr., George Zigre, and Paul Weeks; (second row) Lt. Kevin O'Grady, firefighters Anthony Ambrosio III, Eugene Enfield Jr., Kevin Hopkins, Michael Maurath, Robert Torres, Brian Stauder, William Doherty, John Lindquist Jr., Brian Szebenyi, Thomas Shjarback,

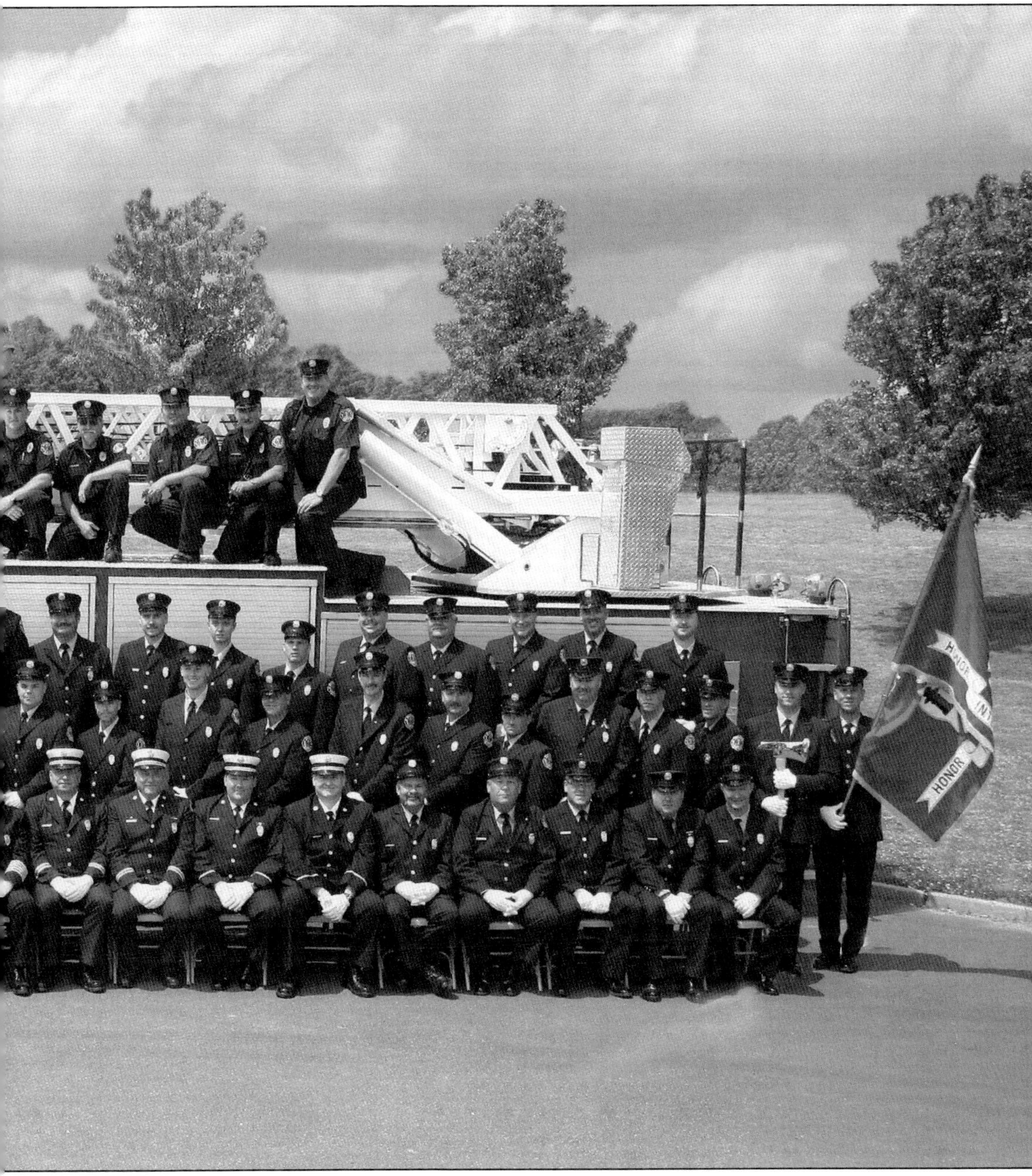

David Gordano, James Walsh, Milton Asprocolas, Matthew Colletto, Scott Disbrow, John McGowan, Robert Sawicki, Glenn Hunter, Kenneth Zawrotniak, Robert Mellinger, Richard Ball, Daniel Maurath, Brian Schreck, Patrick J. Novia, and Timothy Castleton; (third row) firefighters Michael Sacchi, Francis Eosso, Bruce Gregory, Allan Yourstone Jr., Brian Daugherty, Alex Naiduk Jr., Richard Schreck, Douglas MacMahon, Andrew Drebych, Jerry Pepin, Brien Oughton, Brian Mulhearn, Michael Duhigg, Ronald Shalonis, Joseph Czech, Jeffrey Coleman, Wayne Enoch, and Philip McManus; (fourth row) firefighters Michael Pellegrino, John Canfield, Donald Gunthner, Lt. Duane Borwegen, firefighters Thomas Aszman, Duane Clause, Robert Mason Jr., Scott McCloughan, and Daniel DiNuzzo. (Courtesy of P.M. Studios.)

Arthur Latham
William Fisher
Harold Drake
Edward Monaghan
Joseph Sovart
Donald Dudics
Calvin Latham
John Galambos
William Schneider
George Zigre
Milton Asprocolas
George Ellmyer
James Burke
Robert Grand-Jean
John Onder
Peter Borwegen
Robert Meluski
Richard Van Sickle
H. Ray Vliet
Francis Colletto
Edward Costello
Harold McGorvin
Charles Grand-Jean
Arthur Harmon
Edward Kashtock
Alfred Milcsik
John Lindquist
Thomas Dalton
Frank Brogan
James Colletto
Dominick DeFrank
Steven Mozsgae
Julius Yuhas
Richard Latham
John Smith, Jr.
Curtis Clark, Jr.
Richard Borwegen
Anthony Colletto
Alexander Danish
John Dublanyk
Robert Kearston
James Metz
Richard Rogan
Joseph Stumpf
David Jensen
Robert Coleman
Roger Vroom
Joseph Horvath, Sr.
Edward Madger
Louis Sands
Charles Toth
Robert Kopac
Robert Lindquist
Thomas Blanchard
Henry Buergel
Donald Freeman
Allan Yourstone
Cashmeir Piedgon
William Keefe
Eugene Berlinski
Richard Campbell
Leon Mazur
Ralph Ambrosio, Jr.
Carl Demko
Daniel Jordan
Richard Sheridan
John Renner
Ralph Banks
Herbert Eayres
G. Robert Campbell
Wesley Latham
John Spitler
James Montanye
Robert Stramara
James Melynak
Fredrick Vickery
Rudolph Asprocolas
Ronald Kermes
Harold Peach
Paul Matulewicz
Joseph Szebenyi
Walter Ulrich
Charles Spearnock
Emery Tibok
Edward Guarnieri
Bruce Almquist
Robert Yackel
Keith McGorvin
Robert Horvath
Patrick F. Novia
Carl Wood
Peter Borwegen
Lane Benson
Wayne Stryker
Albert Lamkie
Michael Costello
David Davis
Joseph Marino
Timothy Asprocolas
Thomas Guarnieri
Peter Anselmo
Eugene Whitecavage
George Taylor, Jr.
Henry Tempe
Raymond Volker
Robert Walsh
Norman Jensen
Robert Meany
Edward Koehler
Milton Asprocolas
Robert LaCour
Robert Sofield
Kenneth Sovart
Patrick Leonard
William Enoch
Daniel Barlics
David Gordano
Douglas Kosup
Peter McElroy
Jack Campbell
Richard Pizzia
Donald Gunthner
Jeff Voorhees
Denis Phalen
John Douglas, Jr.
Paul Matulawicz
Bruce Gregory
John Lindquist, Jr.
Charles Lynch
Mark Anacker
Brian Daugherty
Brian Latham
Albert Stevens
Joseph Szebenyi, III
Steven Bardecker
Edward Grandjean
James Hook
Joseph Horvath
Frank Imbriacco
Phillip Kilijanski
Kevin Latham
David Milcsik
Jerry Pepin
Christian Siech
Thomas Walsh
Bruce Wegel
George Zigre
Gregg Alleman
Richard Aszman
Charles Copperthwaite
Thomas Droppa
Louis Gurdon
Gregory Gush
Eric Johnson
Anthon Landi, III
Scott McCloughan
Kenneth Milcsik
Gregory Schaefer
Thomas Shjarback
Michael Yock
Gilbert Sharp
Duane Clause
Joseph Duffy
Scott Boland
Jeffrey Coleman
Steven Kucinski, Jr.
Raymond Tibok
Robert Comello
Duane Borwegen
Stephen Deak, Jr.
Richard P. Campbell
Kevin O'Grady
David Bell
Glenn Hunter
Scott Law
Joseph Ulrich, Jr.
Joseph Dipple
Matthew Herka
Richard Scheck
Edward Hummel, Jr.
Christopher Wadiak
Michael Pellegrino
Kenneth Zawrotniak
Douglas MacMahon
Brian Schreck
Michael Paelello
Anthony Pepe
James Brennan
Andrew Drebych
Ralph Ambrosio, III
Thomas Aszman
Ronald Shalonis
Robert Doggett
James Gross
Michael Wheeler
John Palmer
Daniel Wood
Eugene Enfield, Jr.
Robert McDonnell
Richard Ball
Francis Hohwald
Christopher Kasperski
John Canfield
Timothy Eosso
David Edelstein
Francis Eosso
Philip McManus
Michael Duhigg
James Walsh
Joseph Anselmo
Michael Sacchi
Patrick J. Novia
Kevin Hopkins
Samuel Kirkpatrick
Paul Toth
Wayne Enoch
Michael Maurath
William Pellegrino
Allan Yourstone, Jr.
Herbert W. Eayres
Jared Schulman
John Luminiello, Jr.
William Doherty
Brian Stauder
Salvatore Datillo
Robert Sawicki
Brian Mulhearn
Anthony Vicidonimi
Scott Reisz
Timothy Castleton
Andrew Toth
Robert Torres
Alex Naiduk, Jr.
John Gorman
Brian Vicidomini
Daniel Maurath
John Kantra
Robert Mellinger
Peter Yackel
Brien Oughton
Michael Banbor
Paul Weeks
Robert Mason, Jr.
Matthew Colletto
Daniel DiNuzzo
Joseph Toth
Anthony Robinson
Brian Szebenyi
John McGowan
Richard Rodriguez
Scott Disbrow

Listed above are all of the career firefighters who have served within the township of Edison.

Seven

Call the Fire Department

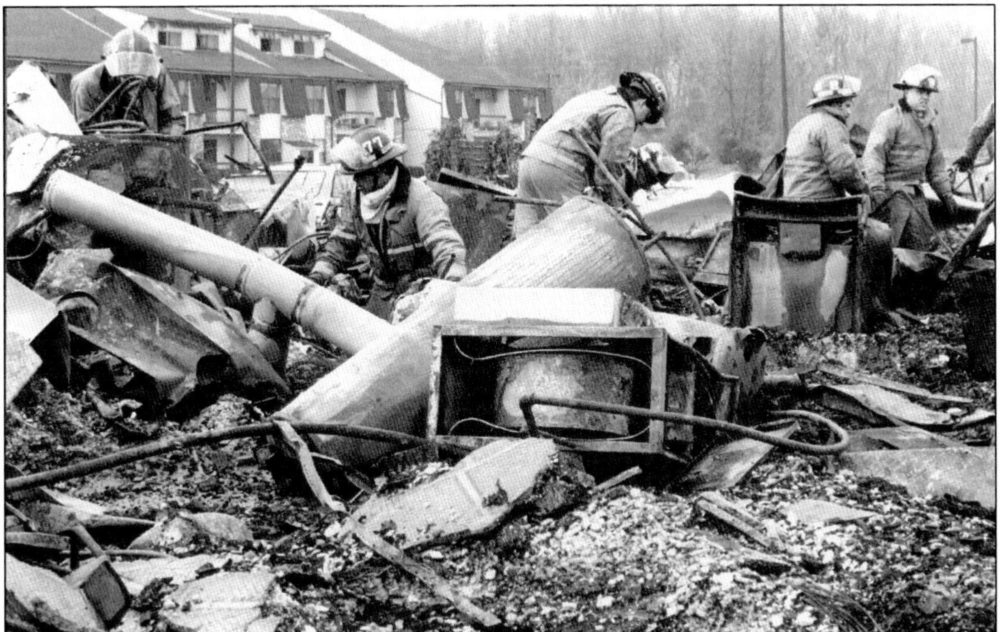

Firefighters are searching through the rubble several hours after the Texas Eastern/Durham Wood gas line explosion that occurred just before midnight on March 23, 1994. This disaster was so intense that the Edison Communication Center received over 200,000 emergency calls within the first hour. An American Airlines pilot flying a passenger jet overhead at the time was later quoted as saying that he thought a nuclear bomb had been dropped on the city of Newark. The blast and the flame from this event were so great that fire departments from up to 50 miles away thought the explosion was in their district. (See page 126 for more details.)

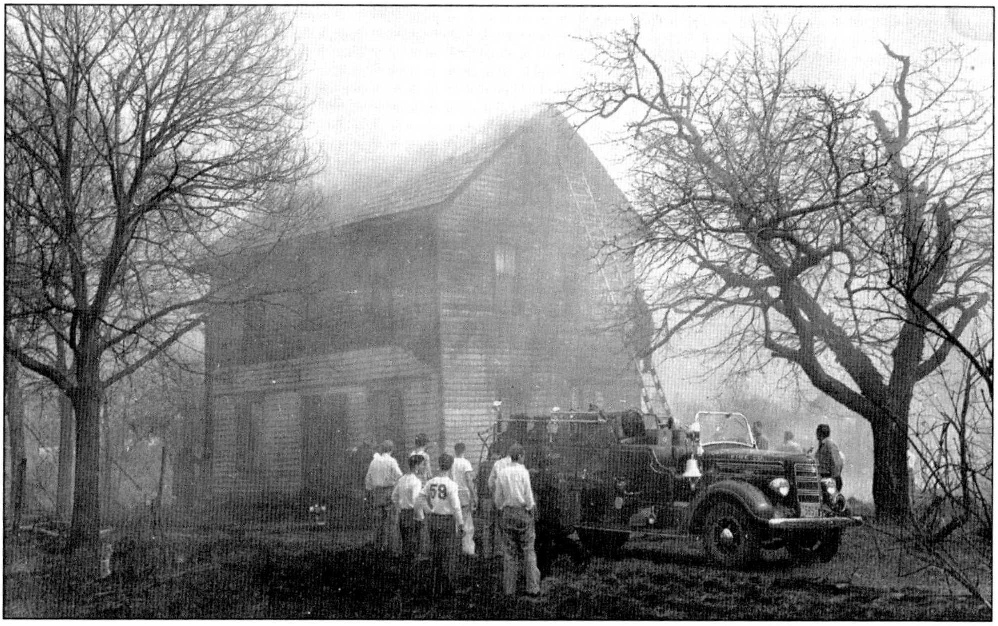

This early photograph shows firefighters from the Oak Tree Volunteer Fire Company fighting a fire at an old Potters farmhouse in the late 1940s. A firefighter ascending a ladder can be seen on the other side of the 1941 Mack fire engine. Potters Crossing was a stop in the Underground Railroad during the Civil War. (Courtesy of Walter Ulrich.)

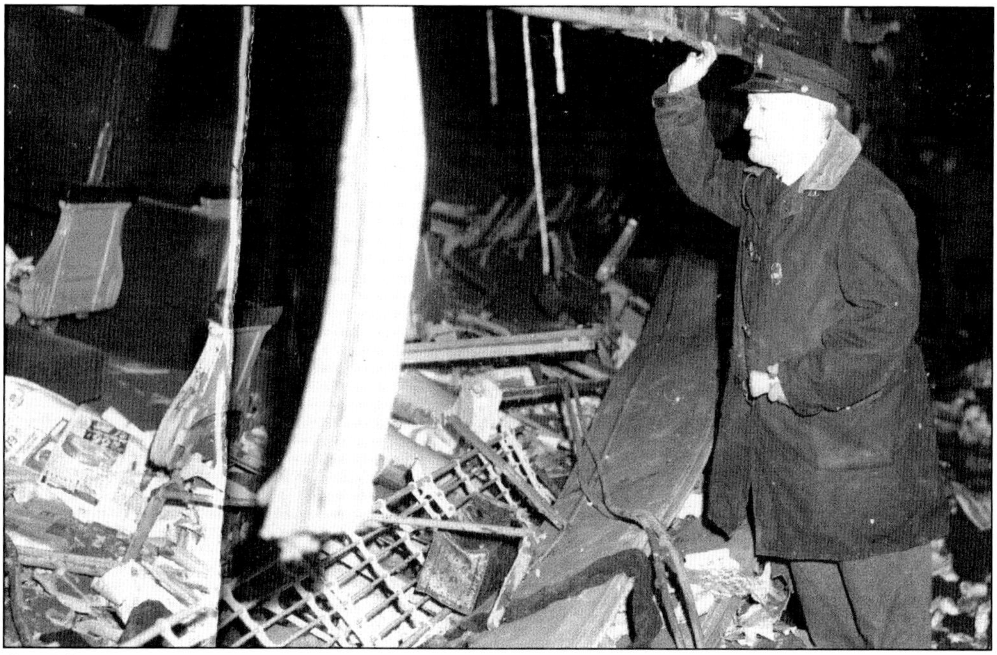

On February 6, 1951, a major train wreck, which resulted in 85 deaths, occurred. This accident took place when a Pennsylvania Railroad commuter train plunged through a temporary overpass in Raritan Township's neighboring town of Woodbridge. Shown above is firefighter Edward Voorhees from Raritan Engine Company No. 1 searching for victims among of the wreckage. (Courtesy of Raritan Engine Company No. 1.)

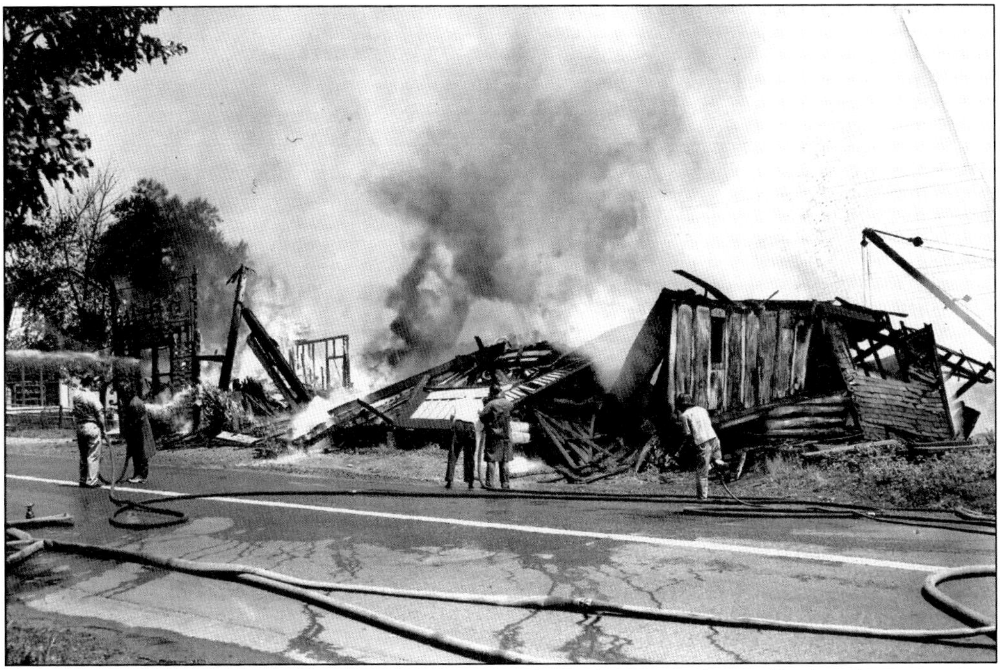

The Jennings Lumber Yard, located on Route 27 along the Pennsylvania Railroad, was a major supplier for the construction of patrol torpedo boats during World War II. On June 6, 1953, Menlo Park firefighters responded to a report of a light smoke condition at the lumber yard. Shortly after their arrival, a dust explosion occurred; which caused the entire structure to become engulfed in flames. It took almost five hours to extinguish the fire. (Courtesy of Edison Volunteer Fire Company No. 1.)

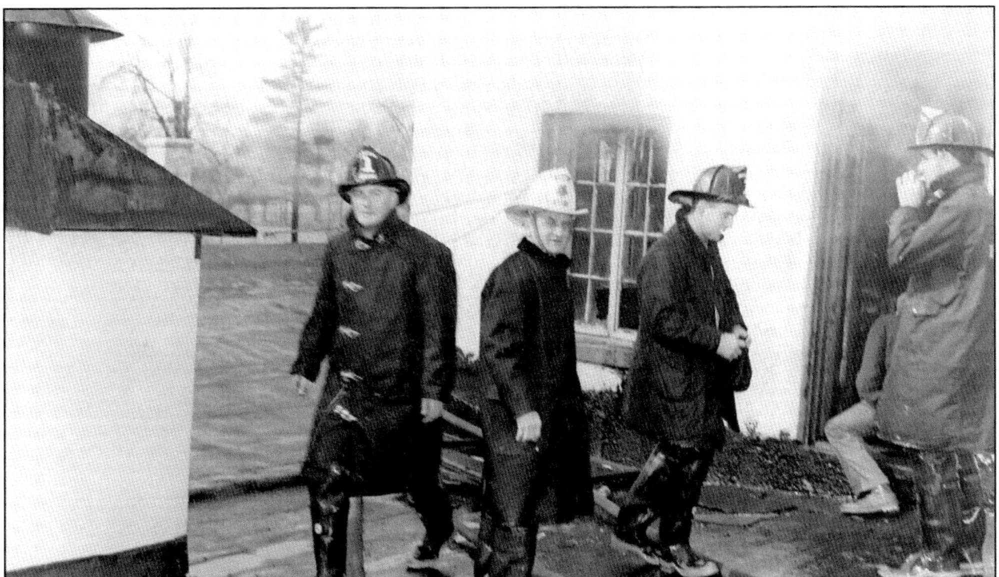

After battling a stubborn kitchen fire at Edison's commonly known Pines Manor in 1956, Chief Engineer Arthur Latham (located in the center wearing a white helmet) is summoned to the roof to assess the damage. (Courtesy Brian Latham.)

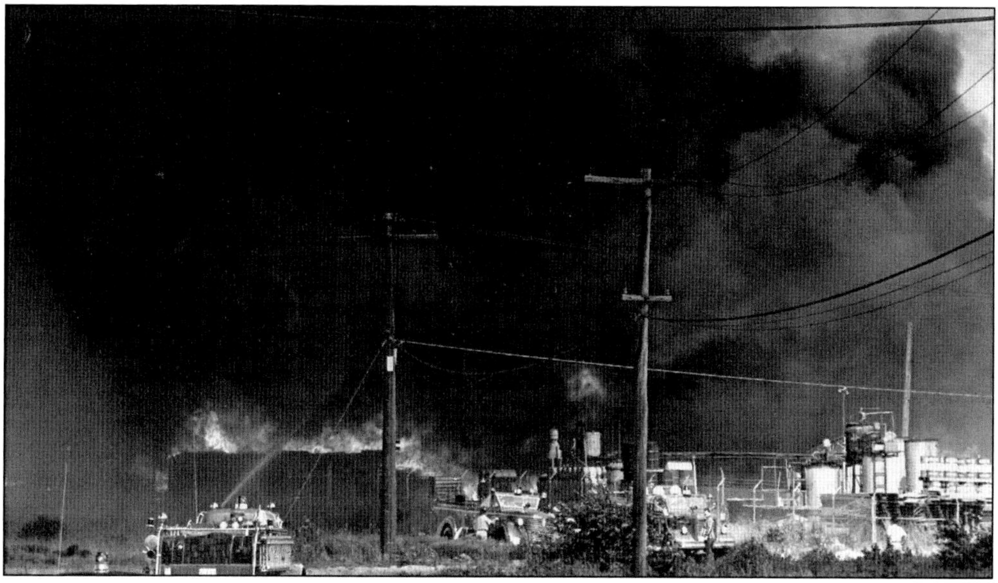

A large chemical fire occurred in the Rotenone Processing Facility at the Chemical Insecticides Corporation on July 3, 1964. The plant, which was located at 30 Whitman Avenue, produced a wide range of insecticides, fungicides, herbicides, and rodenticides between 1958 and 1970. Today the vacant property has been classified as a Superfund Site due to soil contamination. (Courtesy of H. Ray Vliet.).

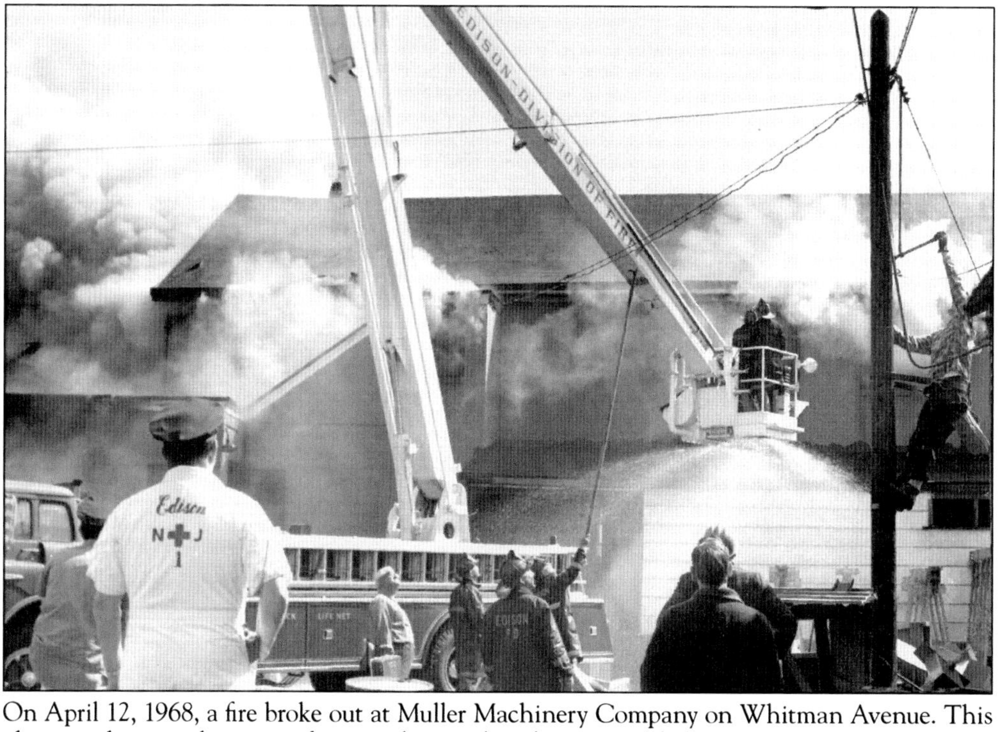

On April 12, 1968, a fire broke out at Muller Machinery Company on Whitman Avenue. This photograph was taken just after a utility worker disconnected the electrical power from the pole so that the firefighters could safely maneuver the Snorkel truck into position. (Courtesy of H. Ray Vliet.)

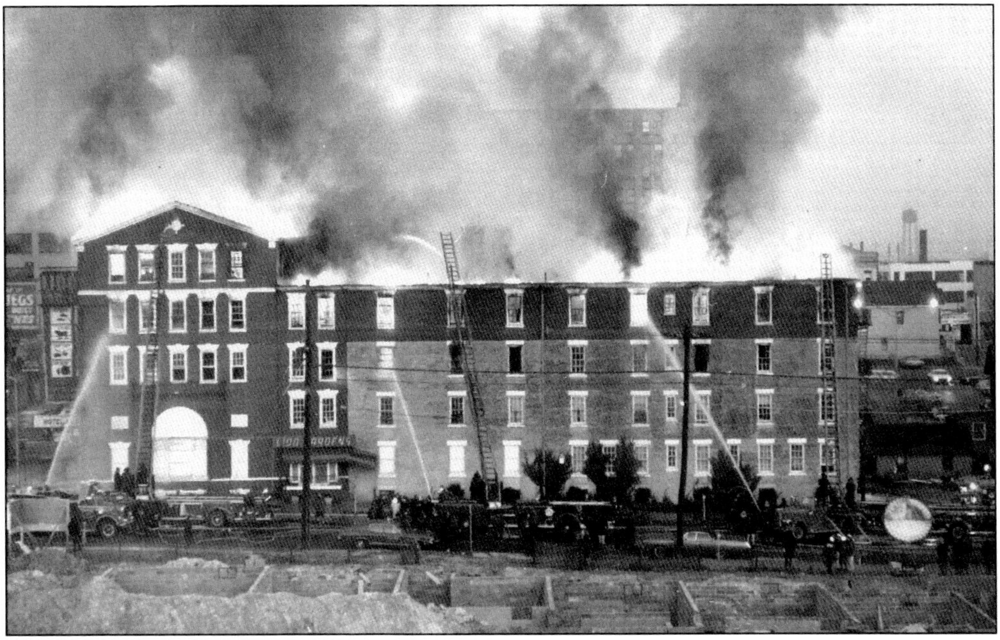

On March 17, 1969, a devastating fire spread through one of Perth Amboy's oldest structures, the Packer House Hotel. The fire, which started on the third floor, claimed the lives of five guests and the hotel manager. Perth Amboy Fire Department promptly called for Edison's Snorkel truck to assist with rescue and extinguishment. (Courtesy of Emery Tibok.)

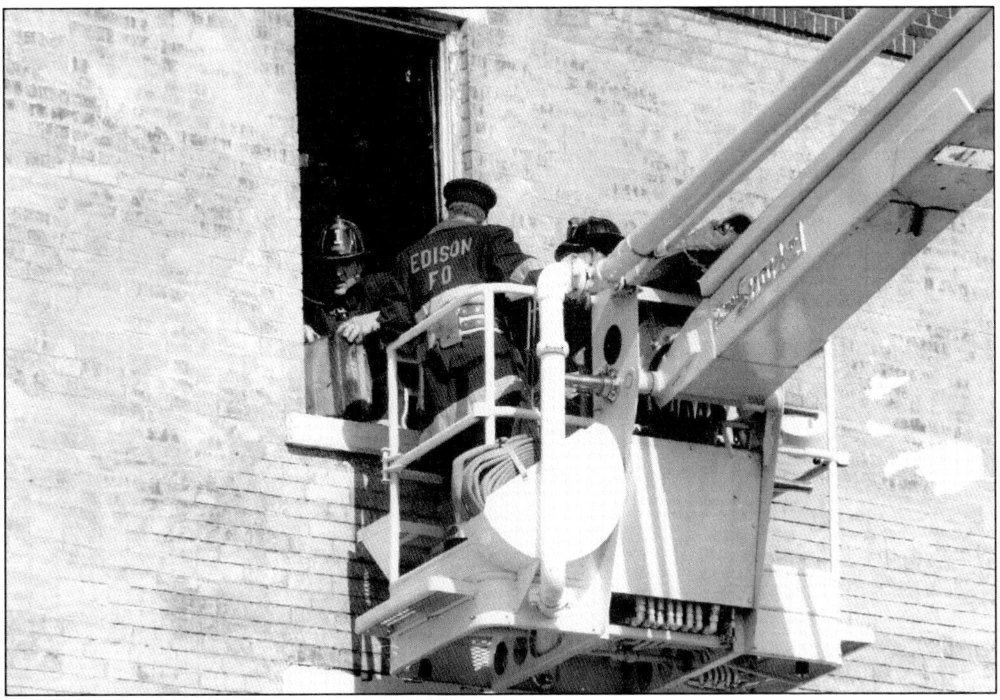

Soon after the extinguishment of the Packer House Hotel fire in Perth Amboy, Edison firefighters Calvin and Richard Latham positioned the Snorkel bucket to assist firefighter Charles Grand-Jean, who was removing a victim from a second floor window. (Courtesy of Emery Tibok.)

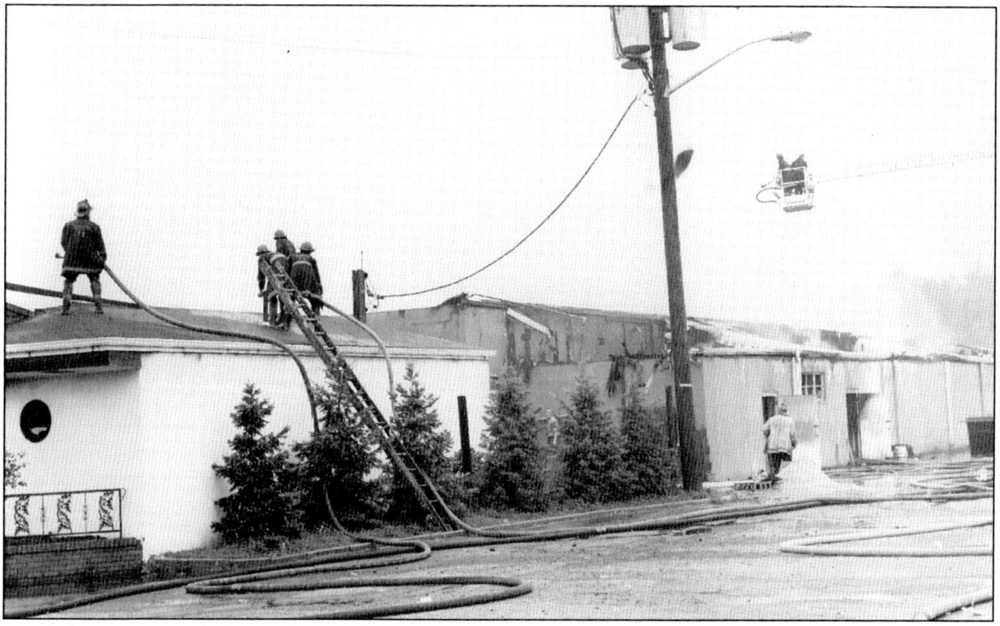

Firefighters responded to a fire at the Royal Oaks Restaurant and Lounge on a hot and muggy morning in August 1970. The restaurant, located at 1700 Oak Tree Road in Edison, sustained heavy structural damage to the ballroom, kitchen, offices, and hallway. The structure was eventually torn down and the Sugar Tree Plaza strip mall was built on the land. (Courtesy of H. Ray Vliet.)

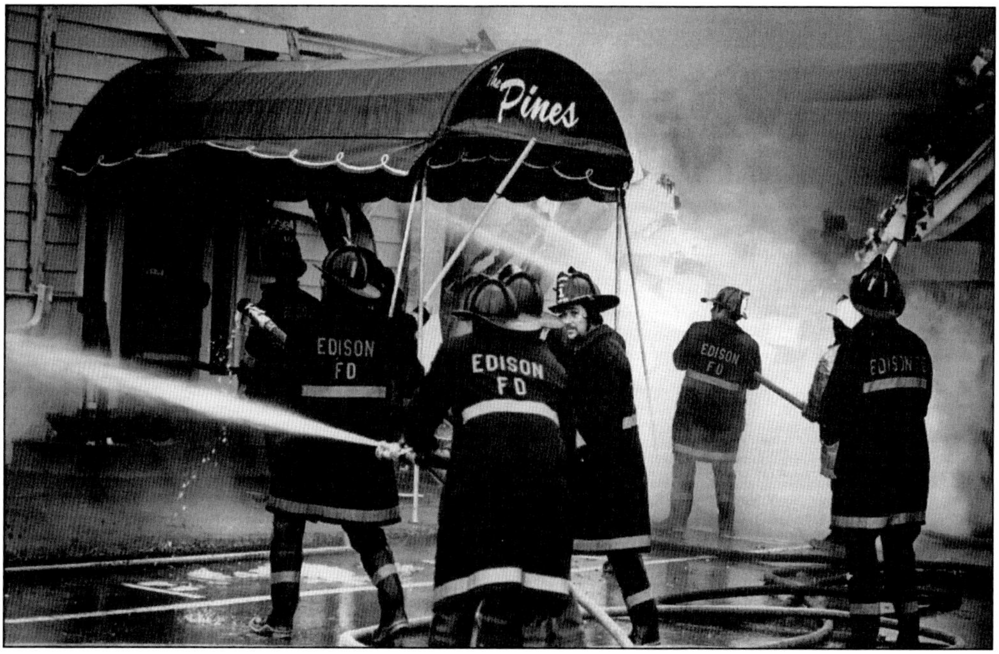

A major fire ripped through Edison's historic landmark catering hall, the Pines Manor, on February 2, 1972. The fire originated in the kitchen and spread throughout the 80-year-old structure. The only occupants at that time were the employees, who escaped without injury. (Courtesy of Edison Division of Fire, photograph by William Gainford.)

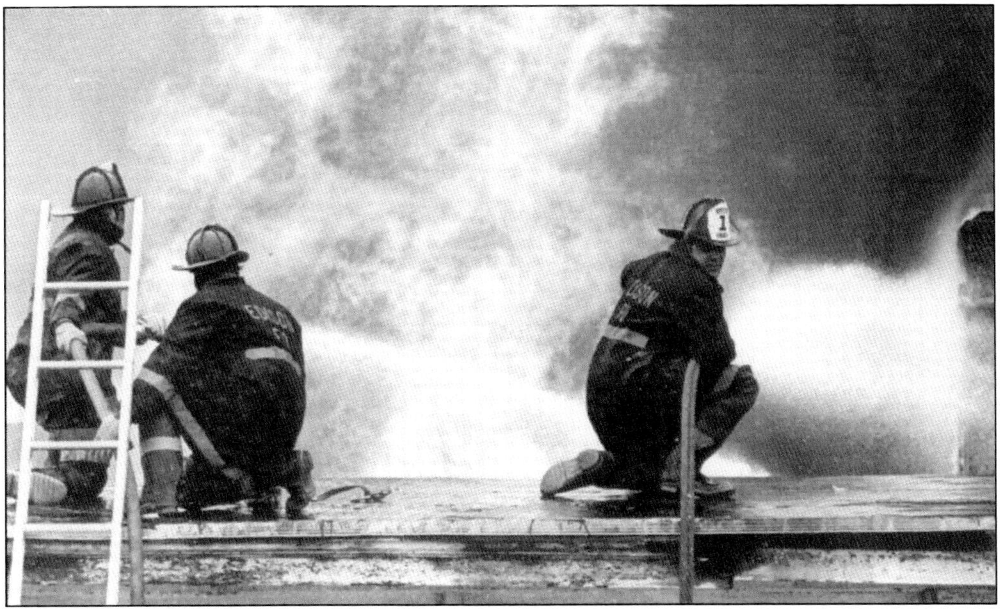

Firefighters, positioned on the roof of the Pines Manor, had a difficult time extinguishing this 1972 fire due to high winds and inadequate water pressure. (Courtesy of Edison Division of Fire, photograph by William Gainford.)

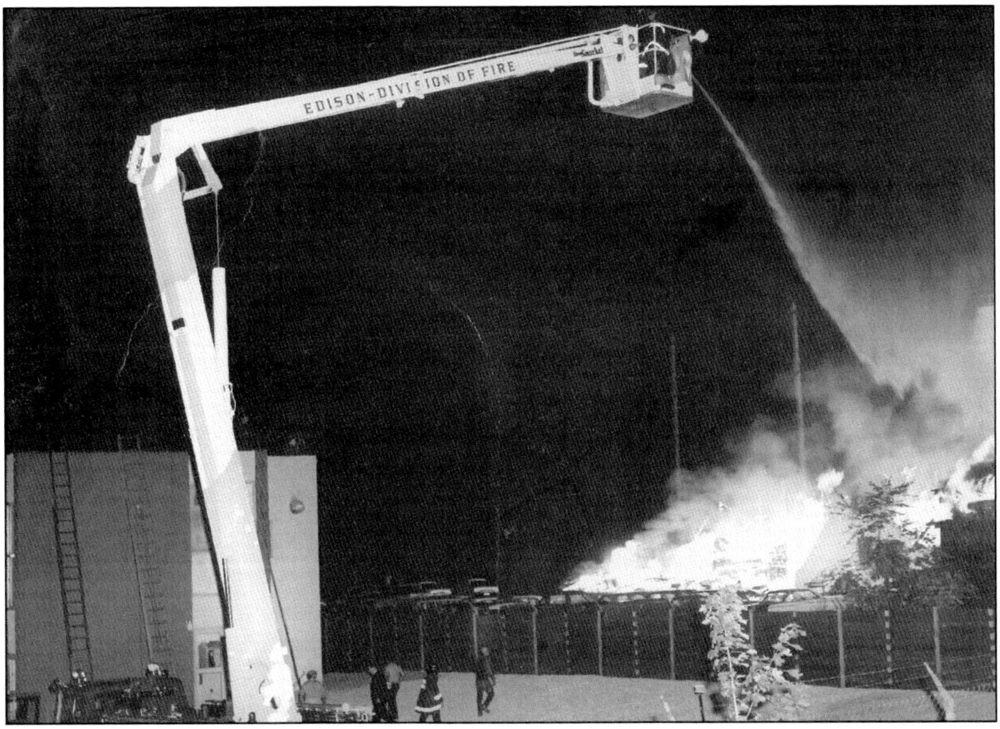

On January 20, 1973, the Edison Division of Fire was called to help extinguish a large fire that spread throughout the H-K Tire Company on Raritan Avenue in neighboring Highland Park. The suspicious fire began around 3:30 a.m. and continued to burn for almost 40 hours. (Courtesy of Edison Division of Fire.)

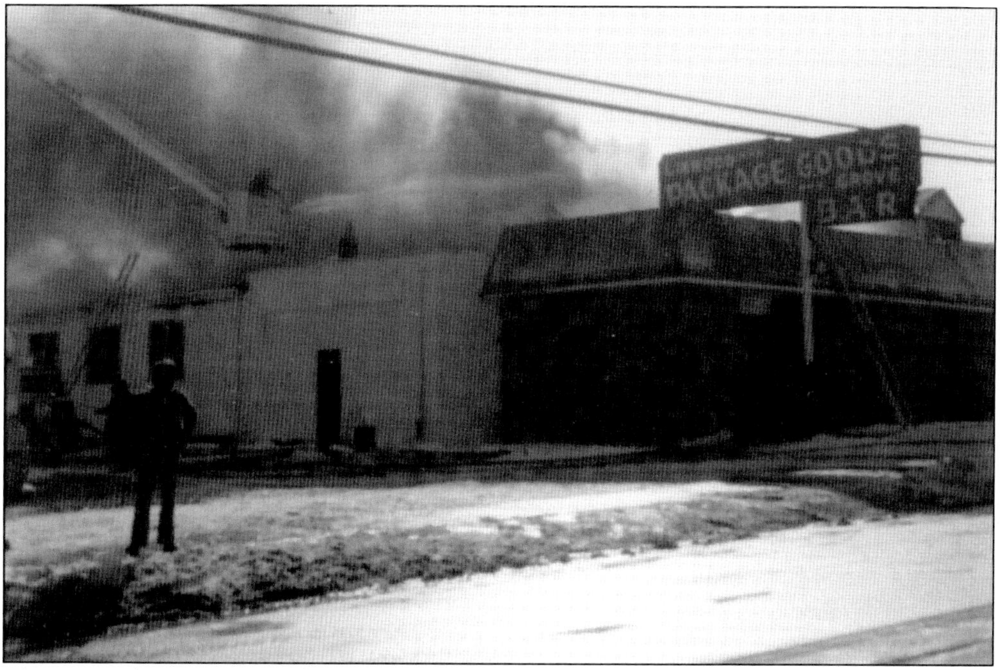

The Linwood Grove Tavern, located at 1876 Route 27 in Edison, was destroyed when a fire broke out in the rear loading dock on December 29, 1976. The tavern, which was purchased in 1947 by Julius and Stephen Kovacs, had been one of Edison's landmarks due to its showcase of popular bands throughout the years. (Courtesy of Richard Campbell.)

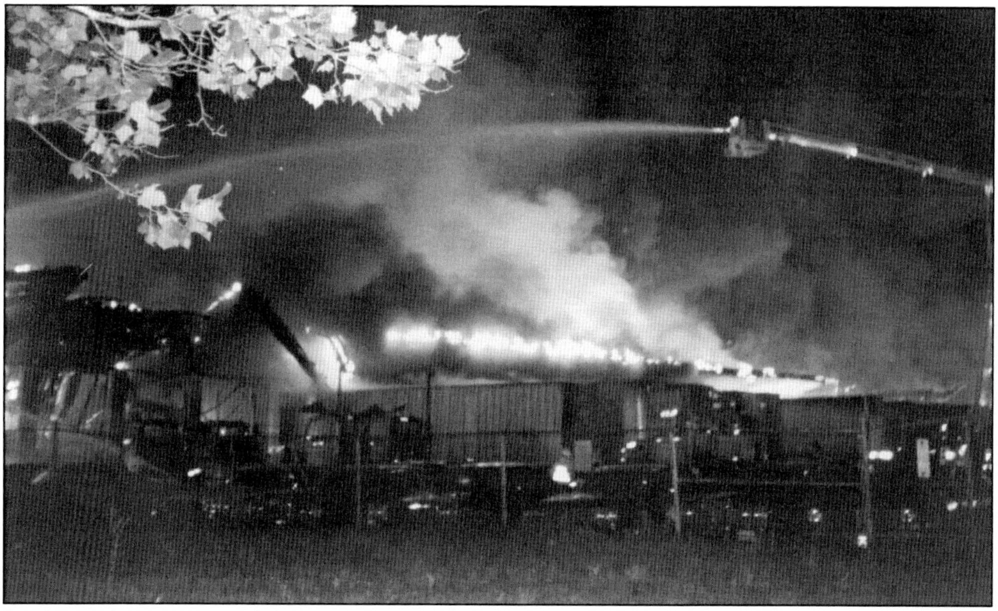

In the early morning of August 3, 1977, firefighters responded to a large fire at the Eastern Trucking Corporation on Route 1 Northbound. Upon arrival, the firefighters found several box trailers lined up against the warehouse fully-involved in fire. In addition to the warehouse and office building, 85 trailers, two tractors, and six passenger vehicles were destroyed in the fire. (Courtesy of William Prairie.)

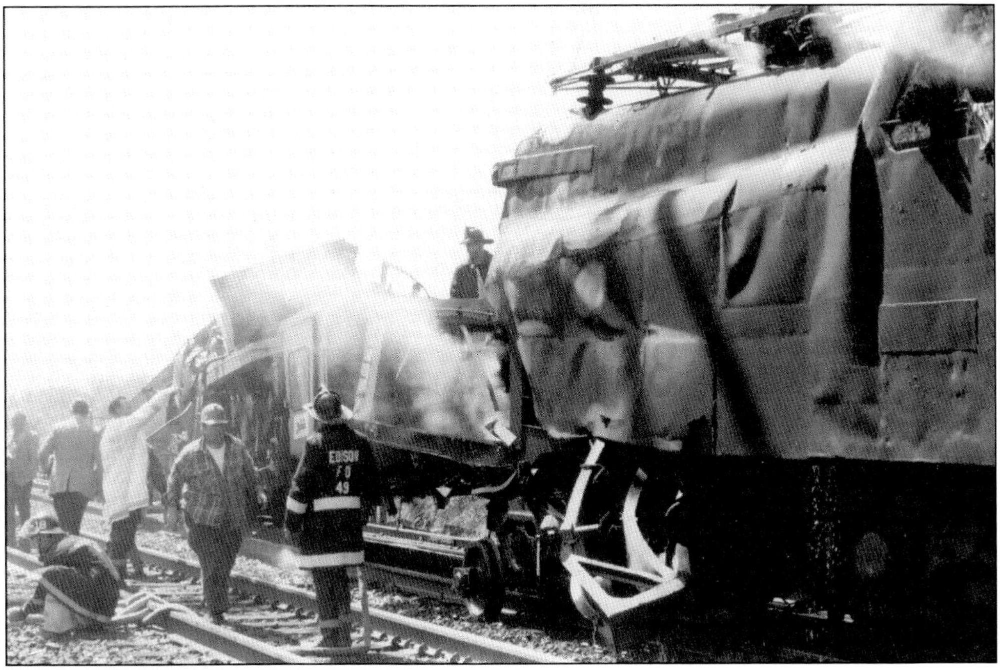

More than 70 passengers suffered minor injuries when a Washington-bound Amtrak train powered by a 4,800 horse-power Class GG-1 locomotive slammed into the rear of a rail work car about one mile south of the Edison train station on April 20, 1979. Deputy Chief Donald Dudics (wearing a white turn-out coat) directs firefighters as they assess the situation and attend to the injured. (Courtesy of Edison First Aid Squad No. 1.)

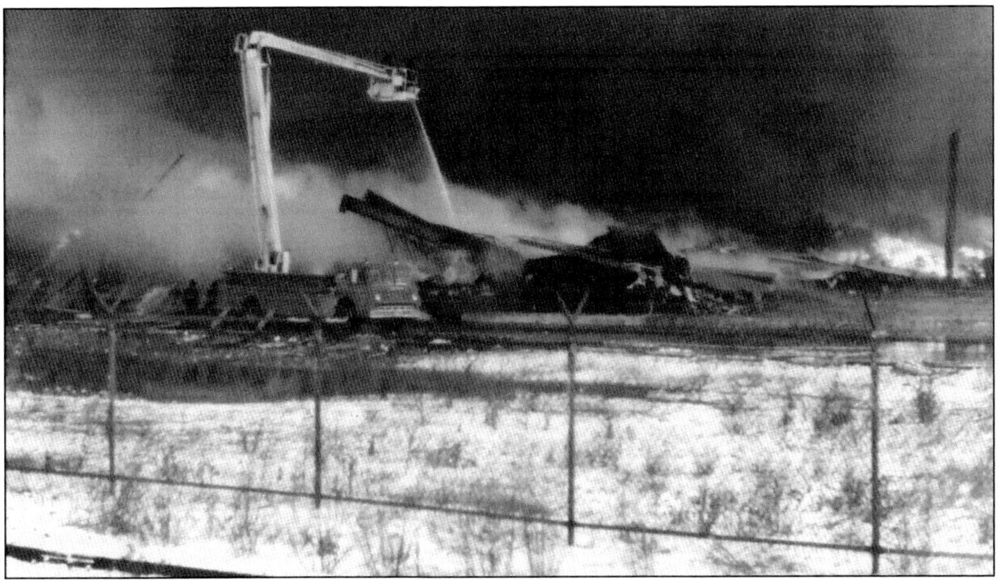

A suspicious fire ripped through the Pathmark Distribution Center off Kilmer Road on January 16, 1979. Even though the 285,000-square-foot warehouse had been outfitted with an automatic sprinkler system, the building was completely destroyed. The photograph above shows the 1961 Ford Snorkel extinguishing the remains after the structure collapsed to the ground. (Courtesy of Edison Division of Fire.)

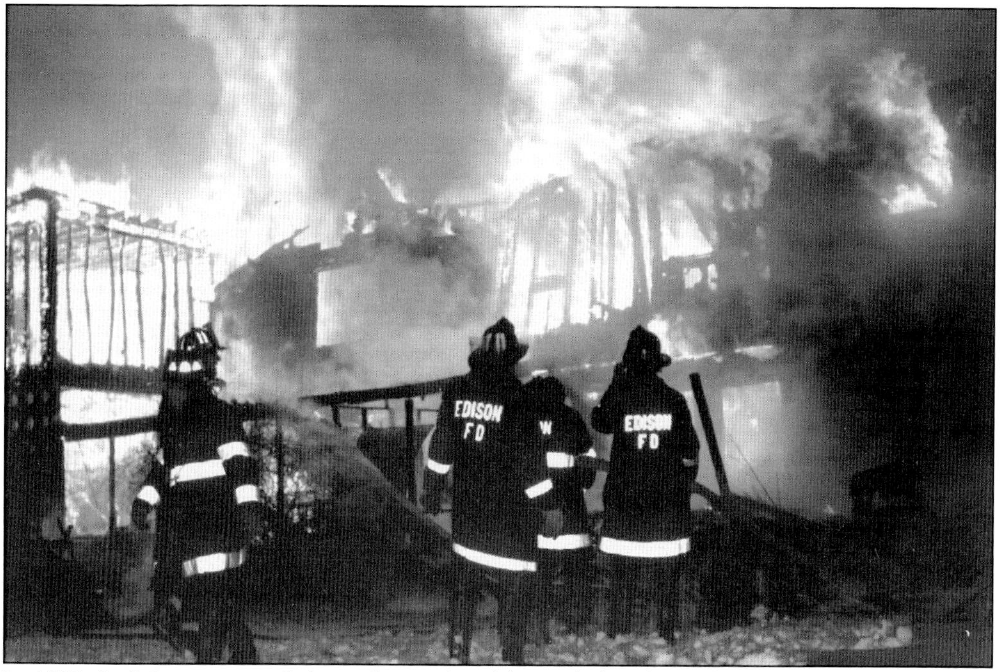

In 1983, firefighters responded to a fire at the old Limoli farmhouse located on Oak Tree Road across from Ventnor Street. This Victorian wood-framed structure, which was built in the late 1800s, became fully engulfed in flame within minutes. It took firefighters several hours to extinguish the blaze on this bone-chilling night. The Jewish Community Center is now on the site of the former Limoli farmhouse. (Courtesy of Richard Aszman.)

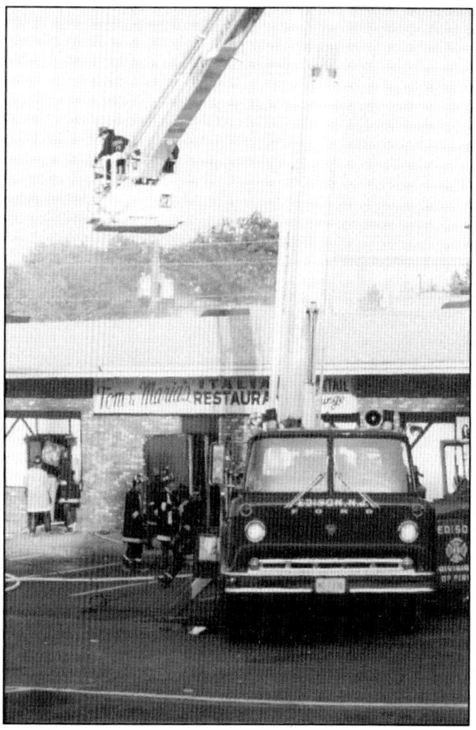

Firefighters Thomas Walsh and Ralph Banks are using the old 1961 Snorkel to access the roof during a fire at Tom and Maria's Italian Restaurant on August 24, 1983. The restaurant, located in the Colonial Village Shopping Plaza on Route 27 and Parsonage Road, sustained extensive damage to the basement and main floor dining area. Fortunately, all employees and patrons exited the establishment unharmed. (Courtesy of Richard Aszman.)

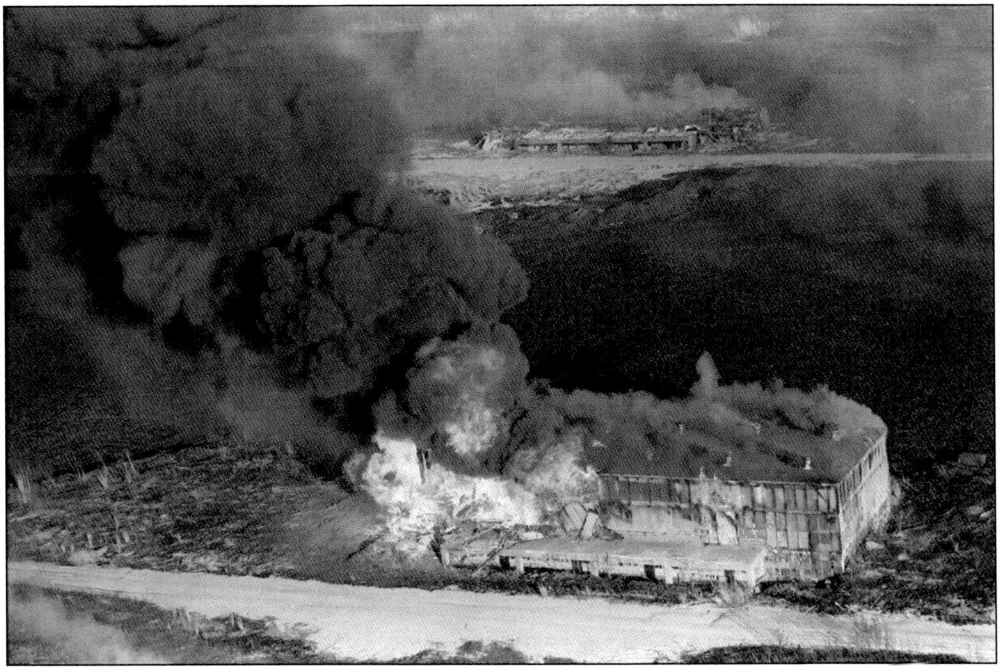

This aerial photograph shows one of 36 buildings that were completely destroyed when a large brush fire swept through the southern portion of Raritan Center Industrial Park on April 30, 1985. Overall, it took the efforts of 17 fire departments and a Coast Guard fire boat to bring the fire under control six hours later. While assisting with command operations, Deputy Fire Chief Arthur Harmon suffered cardiac problems and retired shortly thereafter. (Courtesy of Star Ledger, photograph by Ted Boswell.)

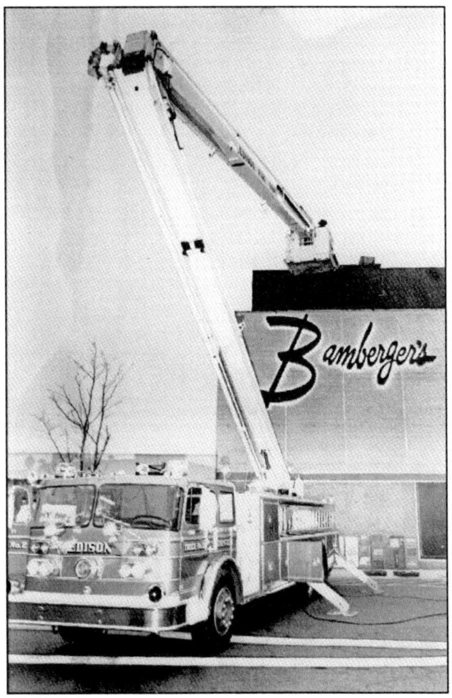

Firefighters from truck 2 are in the process of accessing the roof of Bamberger's department store to check for fire extension during a fire condition on April 20, 1986. Apparently this stubborn fire started in a linen storage area and created a thick smoke condition in the basement. (Courtesy of Jim D'Amico.)

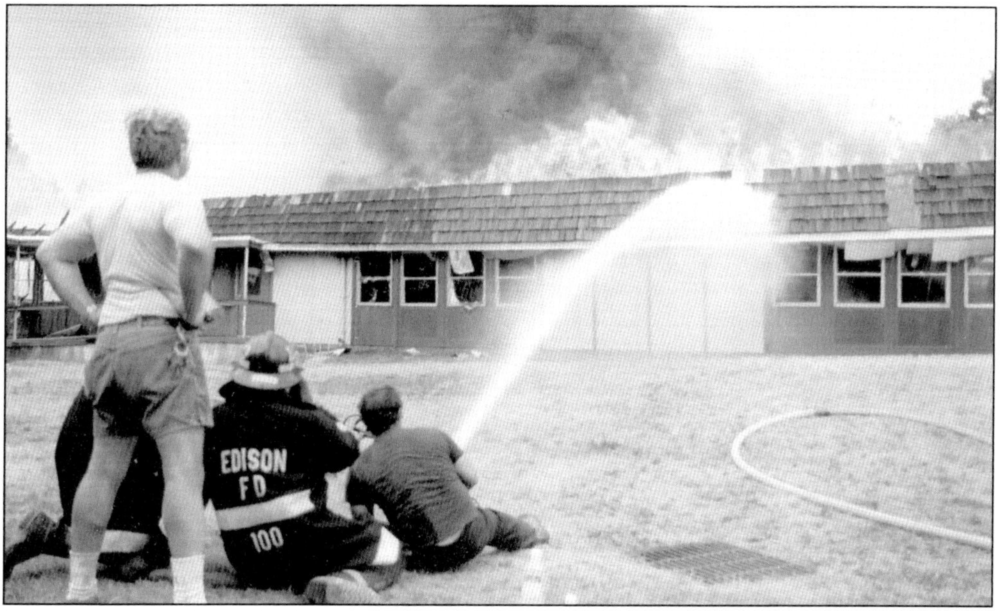

This photograph shows firefighters battling a fire that devastated the Lindeneau School annex on June 27, 1986. The fire started in an evergreen tree next to the building and children were seen fleeing the area just before the structure caught fire. Due to the humid weather and radiant heat from the fire, 10 firefighters were treated for heat exhaustion. The fire was completely extinguished nine hours after the initial alarm. (Courtesy of Jim D'Amico.)

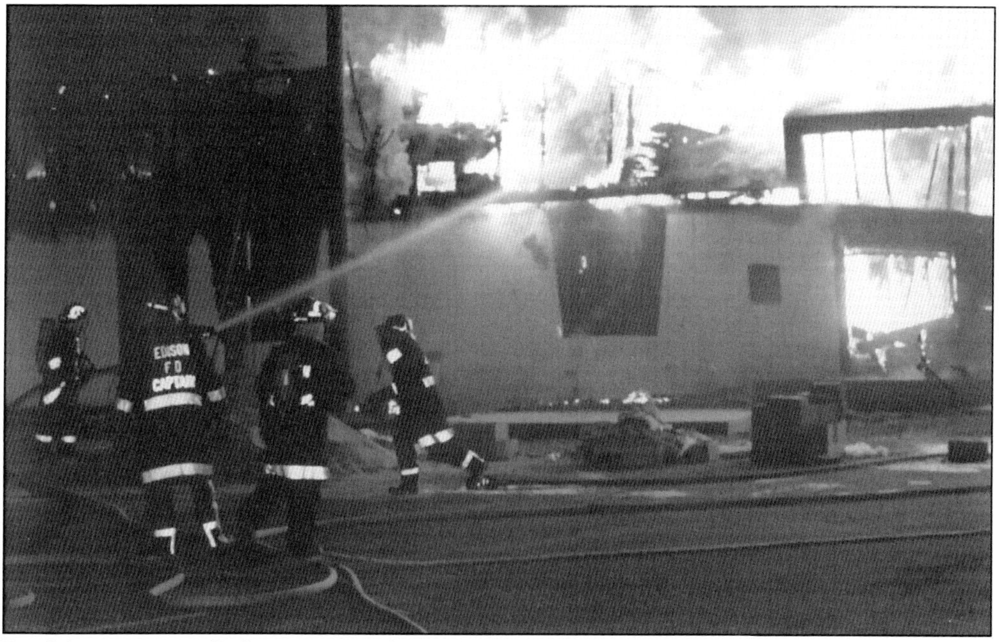

Firefighters were called to extinguish a fully-involved fire, which completely destroyed the Captain's Wheel on March 17, 1988. The vacant restaurant, located on Route 1 South and Wooding Avenue, was in the planning stage to be demolished in order to build a strip mall. Three township juveniles were later arrested for setting the blaze. (Courtesy of Raritan Engine Company No. 1.)

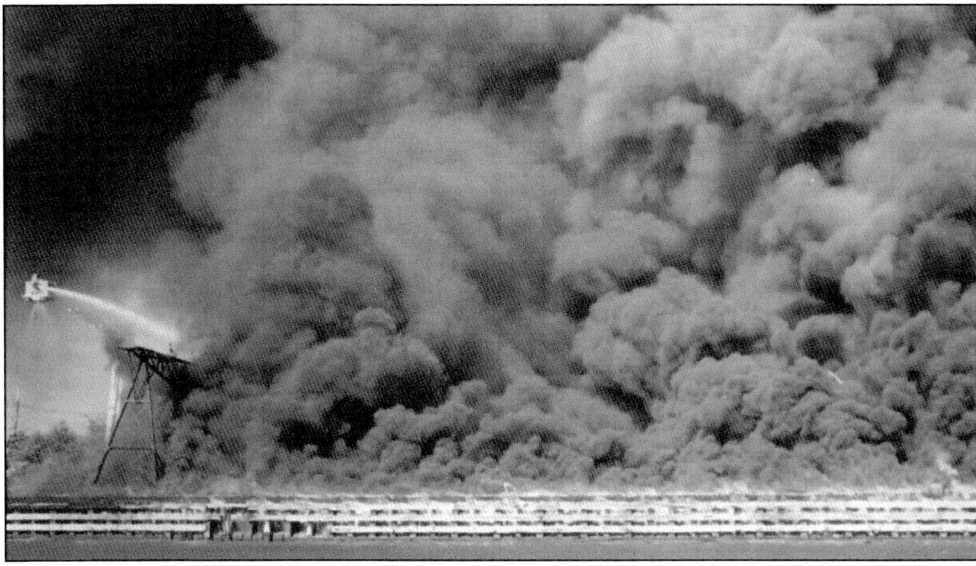

On September 26, 1989, firefighters responded to a fire at the end of Raritan Center Industrial Park. Upon arrival, they were confronted with a very large fire, which was consuming the wooden docks that ran along the Raritan River. Creosol, a chemical used to preserve wood, contributed to the generation of a heavy black cloud of smoke that hovered over the site for hours. Even though firefighters used self-contained breathing apparatus, some suffered chronic respiratory problems that were considered a direct result from this fire. (Courtesy of Edison Division of Fire.)

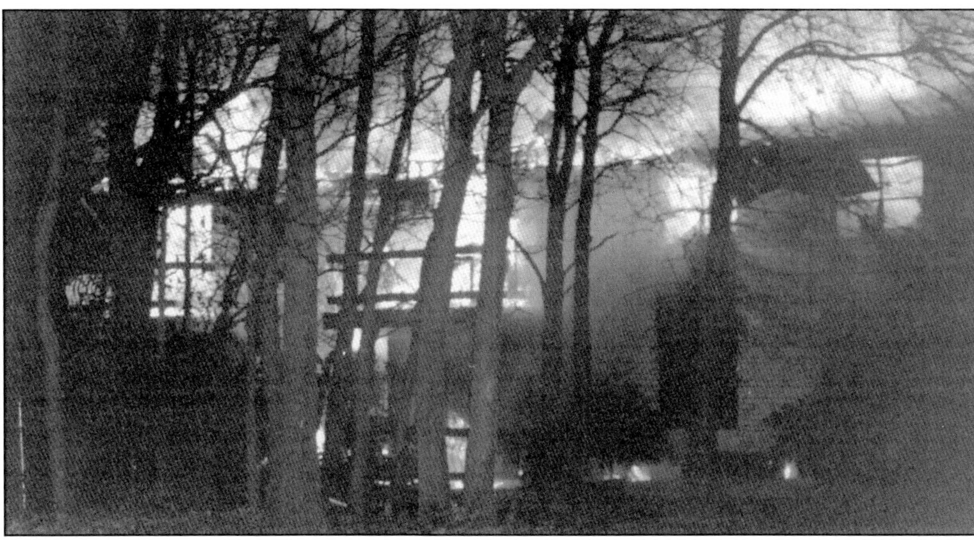

On the morning of December 18, 1989, firefighters responded to an apartment fire on Sunrise Drive. Upon arrival, they found a heavy fire condition on the second floor and a report of a trapped child. Firefighters quickly entered the apartment to conduct a search while an attack line was being stretched. Duane Clause, chief of Raritan Engine Company No. 1, found the nine-year-old child and brought her to safety. While operating the attack line, a firefighter was knocked unconscious from falling debris. Firefighter Anthony Landi III re-entered the structure and removed the firefighter to safety where he received medical attention. (Courtesy of Anthony Landi III.)

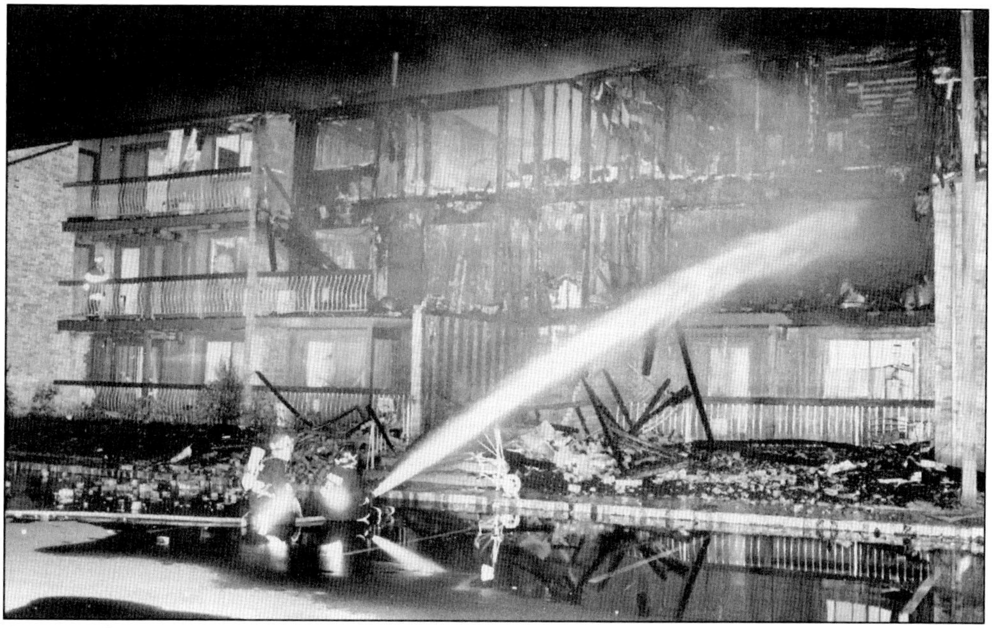

A major fire ripped through building 22 of the Rivendell Apartment Complex on January 9, 1990. The first engine arrived on scene and found a heavy fire condition that was quickly spreading throughout the 16 unit structure. The structure, which was only a few years old, was predominantly constructed of wood that fueled the fire and, eventually, left 31 residents homeless. (Courtesy of Anthony Landi III.)

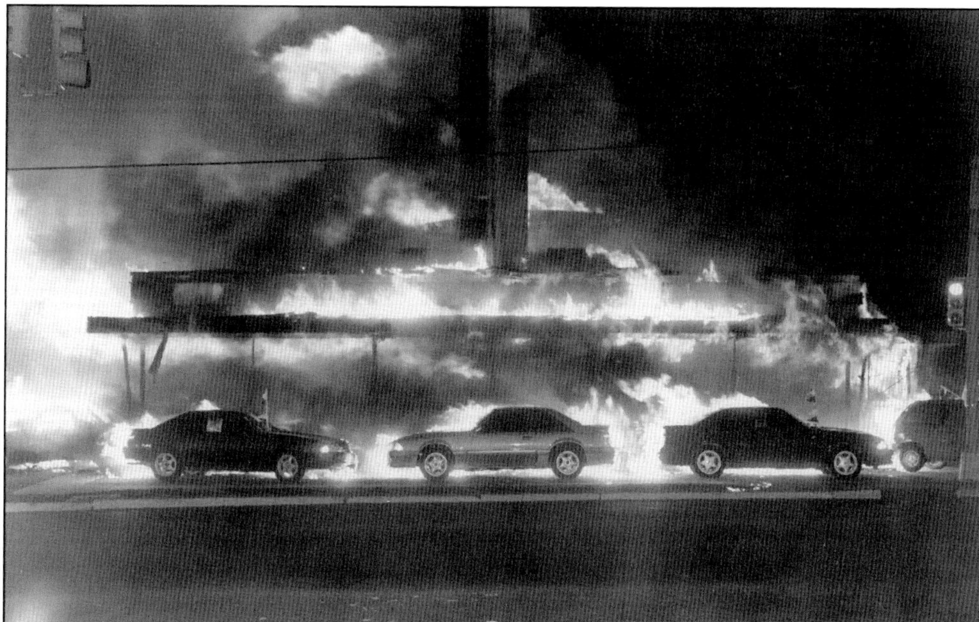

Boro Motors was completely destroyed by fire on the night of February 8, 1991. Even though the dealership was located within the Borough of Metuchen, it had been considered an Edison landmark because it was owned by longtime Edison mayor Anthony Yelencsics. (Courtesy of Jim D'Amico.)

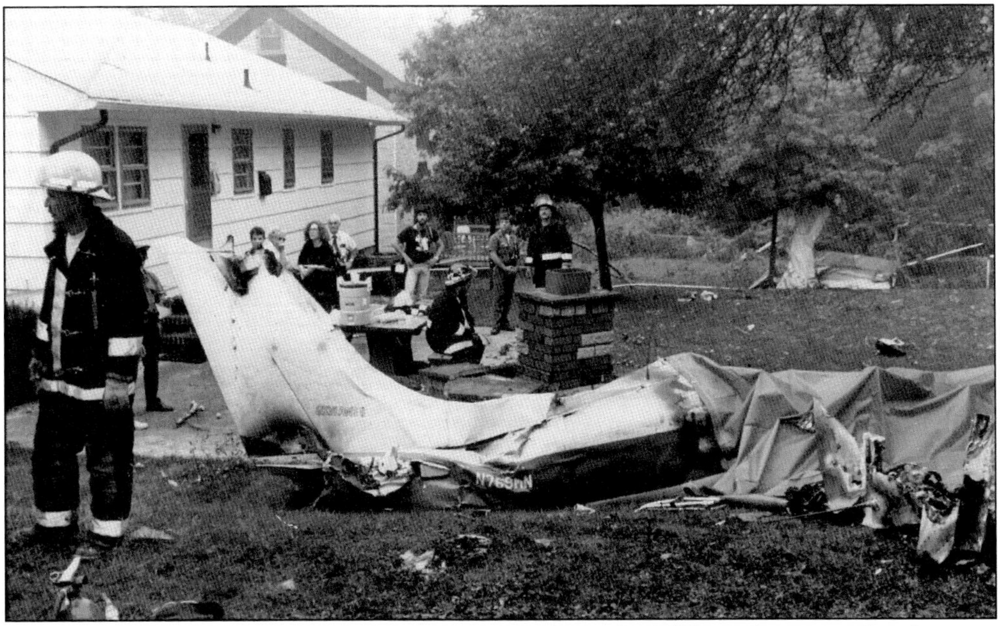

Edison firefighters are no strangers to aviation accidents. In 1955, a pilot escaped with minor injuries after crashing in the woods next to the Pines Manor on Route 27; in 1965, a small plane crashed just off Nixon Lane near Mirror Lake killing two; in 1967, a pilot and student died when a small plane crashed in the backyard of a Jean Place resident; and in 1989, a pilot died after crashing his Cessna airplane in the backyard of a Woodbridge Avenue residence (which is shown above). (Courtesy of Jim D'Amico.)

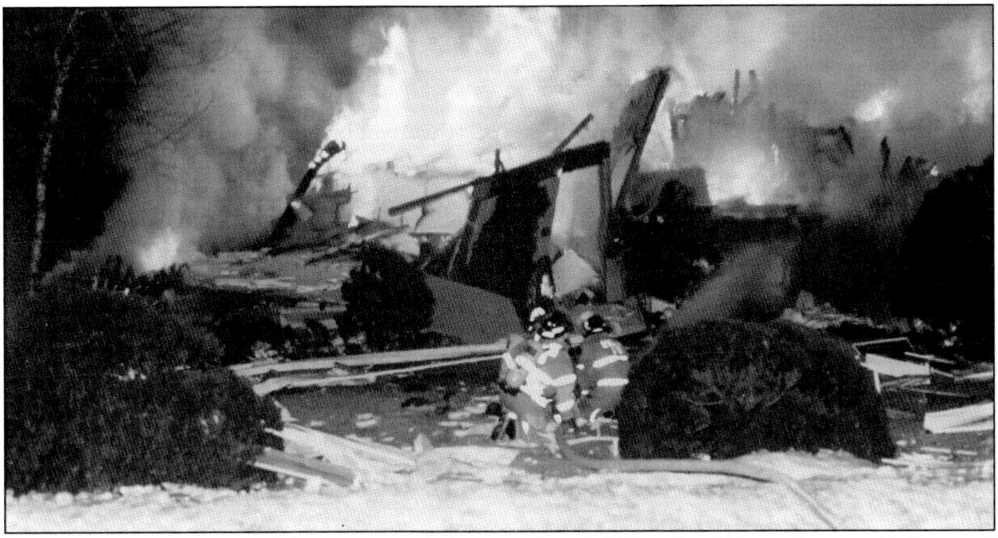

On February 14, 1994, firefighters responded to 288 Plainfield Road in Edison for a smell of natural gas in the residence. The gas company arrived shortly after and released the fire department from the scene. About 30 minutes later, a large explosion occurred and the residence was completely annihilated as flaming debris was spread all over the property. The homeowners, along with their pets, were waiting at a neighbor's house when the blast occurred. This photograph shows firefighters dowsing the flaming piles of rubble minutes after the explosion. (Courtesy of Edison Division of Fire.)

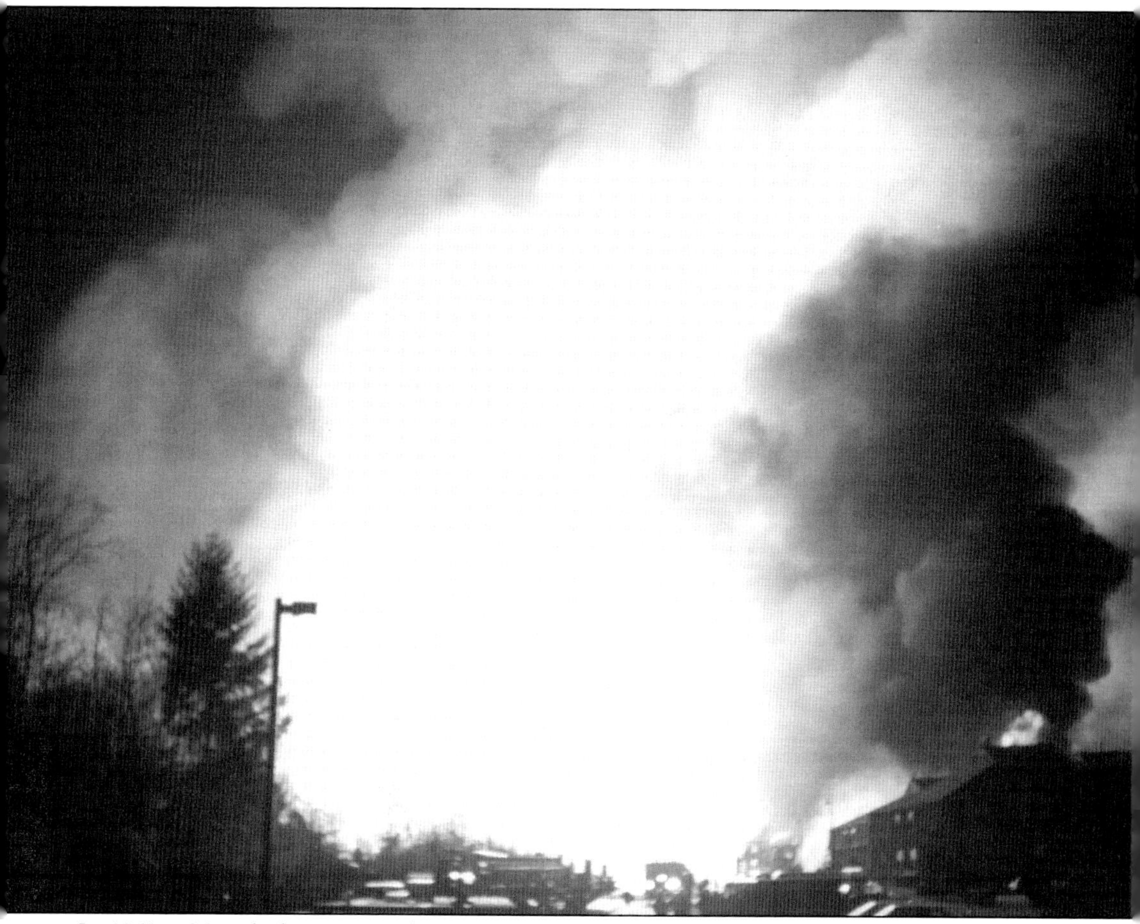

In 1961, a 36-inch-diameter gas line was installed to supply natural gas from the Mexico-Texas border up to a distribution center in Linden, New Jersey, where it branches off before continuing up to the Boston area. This 900-pound-per-square-inch high-pressured line, owned by Texas Eastern Incorporated, runs directly through Edison along a railroad right-of-way. The gas line worked without problems for over 30 years until the night of March 23, 1994. Just minutes before midnight, the pipe ruptured, shook the ground, and gave residents in the nearby Durham Woods Apartment Complex notice to quickly retreat. About three minutes later, the escaping gas found an ignition source and delivered a 500-foot fireball. Overall, it took numerous fire companies from all over to mutually extinguish the apartment buildings, while the gas was being shut off at the closest valve facility. As a result, eight apartment buildings were completely destroyed and eight others were severely damaged. The Texas Eastern gas line explosion will certainly be remembered as the largest explosion and fire Edison firefighters have ever encountered. (Courtesy of Edison Division of Fire.)

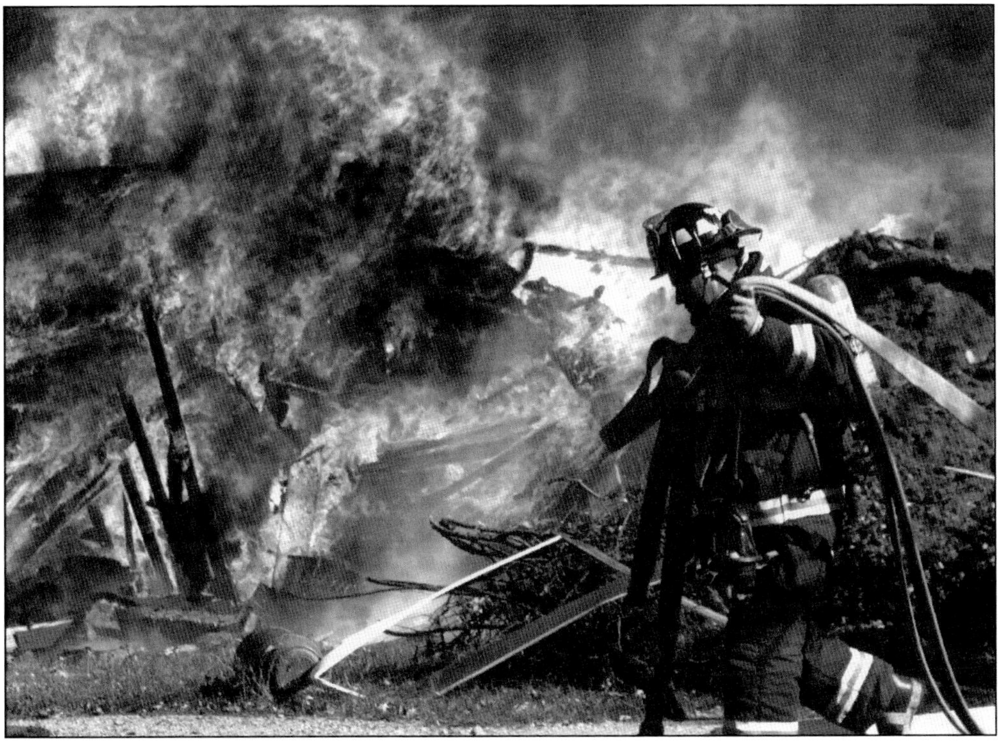

On October 12, 2004, firefighters responded to an explosion at 174 Plainfield Avenue. Upon their arrival, they found the single-family dwelling completely annihilated due to a natural gas explosion with houses on either side burning. This photograph shows firefighter Louis Gurdon Jr. preparing to battle the blaze. (Courtesy of Home News Tribune.)

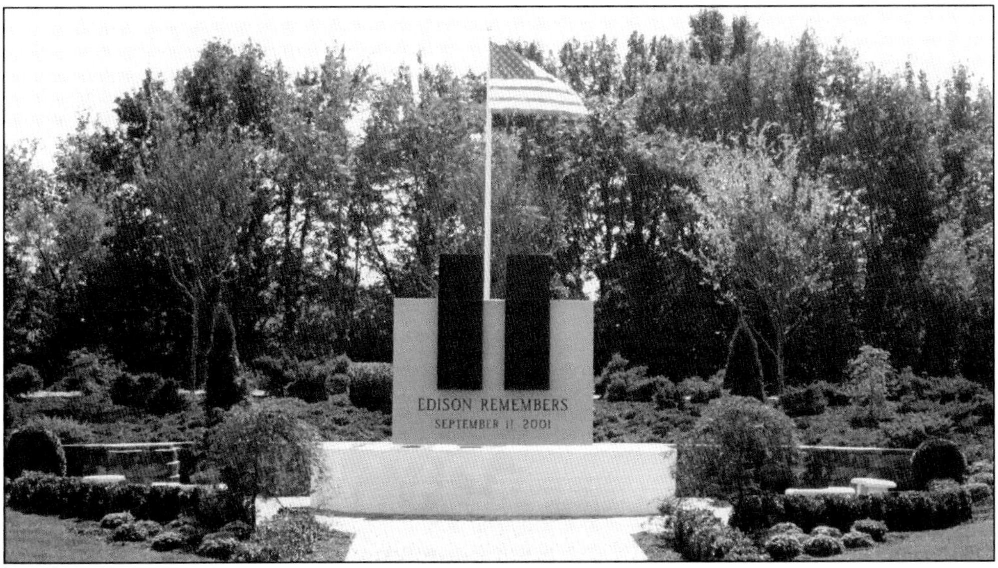

This memorial, which is located near Lake Papianni, honors those who lost their lives on September 11, 2001, and also serves as a constant reminder of the dangers involved in firefighting. Prayers are sent for the safe return of every firefighter at the end of each day and for the loved ones who hope for the same. (Courtesy of Edison Division of Fire.)